Water Technology Management
Challenges and Choices

Ravinder Kumar Kashyap

Oxford Book Company
Jaipur • New Delhi

Published By

Oxford Book Company

Regd. Off. "Bony Residency", Gate No. 2
Opp. Tilak Public School, Vishweswariya Nagar,
Gopalpura Road, Jaipur - 302018, India
Ph: 0141 - 2761280, Telefax: 2761381
email - oxfordbookjpr@gmail.com
website - oxfordbookcompany.com

Delhi Office: 102, 1st Floor, Satyam House,
4327/3, Ansari Road, Darya Ganj,
New Delhi - 110002, India
Ph: 011-45652440

Distributed By

KEYA INTERNATIONAL PRESS LLP

E-mail: keyainternationalpressllp@gmail.com

Published in India, 2024

ISBN : 9788189473679

Price: ₹ 1990.00

Typeset by Shivangi Computers. Jaipur

Printed and bound in India by Thomson Press Ltd.,
New Delhi

Preface

Water is the most common substance on Earth, covering more than 70% of the planet's surface. Yet, the strain of demands of an ever-increasing population has meant that clean, safe water has become a precious commodity, and water supply in many nations, especially in the developing ones, is on issue of immense concern. Competing uses of water for energy, agriculture, and municipal and industrial needs, couples with associated environmental and instream flow requirements (for hydroelectric power, for example) have caused severe local problems. Diminished water pressure, declining spring and steam flow, land subsidence, saltwater intrusion and water pollution are some of these issues.

In such a scenario, water technology management, which incorporates the principles of science and engineering in conserving water resources for drinking and other purposes emerges as a crucial input to maintain and enhance water supply. This text aims to acquaint readers with what entails water technology management, especially as the challenges of water preservation are many, and so are the options associated with it. Analysing current trends and developments, it also brings within purview the insight into the future of the field.

Ravinder Kumar Kashyap

Preface

[illegible] salt-water intrusion and water pollution are some of these issues.

In such a scenario, water technology management, which incorporates the principles of science and engineering in conserving water resources for drinking and other purposes emerges as a crucial input to [illegible] and enhance water supply. This text aims to acquaint readers with what entails water technology management, especially as the challenges of water preservation are many, and so are the options associated with it. Analysing current trends and developments, it also brings within purview the insight into the future of the field.

Kanwaljit Kumar Kashyap

Contents

1

Introduction

Water can appear in three phases. Water takes many different forms on Earth: water vapour and clouds in the sky; seawater and rarely icebergs in the ocean; glaciers and rivers in the mountains; and aquifers in the ground.

Water can dissolve many different substances, giving it different tastes and odors. In fact, humans and other animals have developed senses to be able to evaluate the potability of water: animals generally dislike the taste of salty sea water and the putrid swamps and favour the purer water of a mountain spring or aquifer. Humans also tend to prefer cold water rather than lukewarm, as cold water is likely to contain less microbes. The taste advertised in spring water or mineral water derives from the minerals dissolved in it, as pure H_2O is tasteless. As such, purity in spring and mineral water refers to purity from toxins, pollutants, and microbes.

CHEMICAL AND PHYSICAL PROPERTIES OF WATER

The major chemical and physical properties of water are:

— Water is a tasteless, odourless liquid at ambient temperature and pressure. The colour of water and ice are, intrinsically, a very light blue hue, although water appears colourless in small quantities. Ice also appears colourless, and water vapour is essentially invisible as a gas.

— Water is transparent, and thus aquatic plants can live within the water because sunlight can reach them. Only strong UV light is slightly absorbed.

— Water is a liquid under standard conditions.

— Since oxygen has a higher electronegativity than hydrogen, water is a polar molecule. The oxygen has a slight negative charge while the hydrogens have a slight positive charge giving the article a strong effective dipole moment. The interactions between the different dipoles of each molecule cause a net attraction force associated with water's high amount of surface tension.

— Another very important force that causes the water molecules to stick to one another is the hydrogen bond.

— The boiling point of water (and all other liquids) is directly related to the barometric pressure. For example, on the top of Mt. Everest water boils at about 68 °C (154 °F), compared to 100 °C (212 °F) at sea level. Conversely, water deep in the ocean near geothermal vents can reach temperatures of hundreds of degrees and remain liquid.

— Water sticks to itself. Water has a high surface tension caused by the strong cohesion between water molecules because it is polar. The apparent elasticity caused by surface tension drives the capillary waves.

— Water also has high adhesion properties because of its polar nature.

— Capillary action refers to the tendency of water to move up a narrow tube against the force of gravity.

— Water is a very strong solvent, referred to as the *universal solvent*, dissolving many types of substances. Substances that will mix well and dissolve in water, e.g. salts, sugars, acids, alkalis, and some gases: especially oxygen, carbon dioxide (carbonation), are known as "hydrophilic" (water-loving) substances, while those that do not mix well with water (e.g. fats and oils), are known as "hydrophobic" (water-fearing) substances.

- All the major components in cells (proteins, DNA and polysaccharides) are also dissolved in water.
- Pure water has a low electrical conductivity, but this increases significantly upon solvation of a small amount of ionic material such as sodium chloride.
- Water has the second highest specific heat capacity of any known chemical compound, after ammonia, as well as a high heat of vaporisation (40.65 kJ mol^{-1}), both of which are a result of the extensive hydrogen bonding between its molecules. These two unusual properties allow water to moderate Earth's climate by buffering large fluctuations in temperature.
- The maximum density of water is at 3.98 °C (39.16 °F). Water becomes even less dense upon freezing, expanding 9%. This causes an unusual phenomenon: ice floats upon water, and so water organisms can live inside a partly frozen pond because the water on the bottom has a temperature of around 4 °C (39 °F).
- Water is miscible with many liquids, for example ethanol, in all proportions, forming a single homogeneous liquid. On the other hand, water and most oils are immiscible usually forming layers according to increasing density from the top. As a gas, water vapour is completely miscible with air.
- Water forms an azeotrope with many other solvents.

ADR label for transporting goods dangerously reactive with water:

- Some substances (sodium, lithium, calcium, potassium) emit a flammable gas (hydrogen) when wet, or react violently with water.

DISTRIBUTION OF WATER IN NATURE

Water in the Universe

Much of the universe's water may be produced as a byproduct of star formation. When stars are born, their birth is accompanied by a strong outward wind of gas and dust. When this outflow of material eventually impacts the surrounding gas, the shock waves that are created compress and heat the gas. The water observed is quickly produced in this warm dense gas.

Water has been detected in interstellar clouds within our galaxy, the Milky Way. It is believed that water exists in abundance in other galaxies too, because its components, hydrogen and oxygen, are among the most abundant elements in the universe. Interstellar clouds eventually condense into solar nebulae and solar systems, such as ours. Water covers 71% of the Earth's surface; the oceans contain 97.2% of the Earth's water. The Antarctic ice sheet, which contains 90% of all fresh water on Earth, is visible at the bottom. Condensed atmospheric water can be seen as clouds, contributing to the Earth's albedo.

Hydrology is the study of the movement, distribution, and quality of water throughout the Earth. The study of the distribution of water is hydrography. The study of the distribution and movement of groundwater is hydrogeology, of glaciers is glaciology, of inland waters is limnology and distribution of oceans is oceanography. Ecological processes with hydrology are in focus of ecohydrology.

The collective mass of water found on, under, and over the surface of a planet is called hydrosphere. Earth's approximate water volume (the total water supply of the world) is 1,360,000,000 km^3 (326,000,000 mi^3). Of this volume:

— 1,320,000,000 km^3 (316,900,000 mi^3 or 97.2%) is in the oceans

— 25,000,000 km^3 (6,000,000 mi^3 or 1.8%) is in glaciers, ice caps and ice sheets.
— 13,000,000 km^3 (3,000,000 mi^3 or 0.9%) is groundwater.
— 250,000 km^3 (60,000 mi^3 or 0.02%) is fresh water in lakes, inland seas, and rivers.
— 13,000 km^3 (3,100 mi^3 or 0.001%) is atmospheric water vapour at any given time.

Groundwater and fresh water are useful or potentially useful to humans as water resources.

Liquid water is found in bodies of water, such as an ocean, sea, lake, river, stream, canal, pond, or puddle. The majority of water on Earth is sea water. Water is also present in the atmosphere in solid, liquid, and vapour phases. It also exists as groundwater in aquifers.

The most important geological processes caused by water are: chemical weathering, water erosion, water sediment transport and sedimentation, mudflows, ice erosion and sedimentation by glacier.

Water Cycle

The water cycle refers to the continuous exchange of water within the hydrosphere, between the atmosphere, soil water, surface water, groundwater, and plants.

Water moves perpetually through each of these regions in the water cycle consisting of following transfer processes:
— evaporation from oceans and other water bodies into the air and transpiration from land plants and animals into air.
— precipitation, from water vapour condensing from the air and falling to earth or ocean.
— runoff from the land usually reaching the sea.

Most water vapour over the oceans returns to the oceans, but winds carry water vapour over land at the same rate as runoff into the sea, about 36 Tt per year. Over land, evaporation and transpiration contribute another 71 Tt per year. Precipitation, at a rate of 107 Tt per year over land, has several forms: most commonly rain, snow, and hail, with some contribution from fog and dew. Condensed water in the air may also refract sunlight to produce rainbows.

Water runoff often collects over watersheds flowing into rivers. A mathematical model used to simulate river or stream flow and calculate water quality parameters is hydrological transport model. Some of water is diverted to irrigation for agriculture. Rivers and seas offer opportunity for travel and commerce. Through erosion, runoff shapes the environment creating river valleys and deltas which provide rich soil and level ground for the establishment of population centers. A flood occurs when an area of land, usually low-lying, is covered with water. It is when a river overflows its banks or flood from the sea. A drought is an extended period of months or years when a region notes a deficiency in its water supply. This occurs when a region receives consistently below average precipitation.

Fresh Water Storage

Some runoff water is trapped for periods, for example in lakes. At high altitude, during winter, and in the far north and south, snow collects in ice caps, snow pack and glaciers. Water also infiltrates the ground and goes into aquifers. This groundwater later flows back to the surface in springs, or more spectacularly in hot springs and geysers. Groundwater is also extracted artificially in wells. This water storage is important, since clean, fresh water is essential to human and other land-based life. In many parts of the world, it is in short supply.

Tides

Tides are the cyclic rising and falling of Earth's ocean surface caused by the tidal forces of the Moon and the Sun acting on the oceans. Tides cause changes in the depth of the marine and estuarine water bodies and produce oscillating currents known as tidal streams. The changing tide produced at a given location is the result of the changing positions of the Moon and Sun relative to the Earth coupled with the effects of Earth rotation and the local bathymetry. The strip of seashore that is submerged at high tide and exposed at low tide, the intertidal zone, is an important ecological product of ocean tides.

EFFECTS ON LIFE

From a biological standpoint, water has many distinct properties that are critical for the proliferation of life that set it apart from other substances. It carries out this role by allowing organic compounds to react in ways that ultimately allow replication. All known forms of life depend on water. Water is vital both as a solvent in which many of the body's solutes dissolve and as an essential part of many metabolic processes within the body.

Metabolism is the sum total of anabolism and catabolism. In anabolism, water is removed from molecules (through energy requiring enzymatic chemical reactions) in order to grow larger molecules (e.g. starches, triglycerides and proteins for storage of fuels and information). In catabolism, water is used to break bonds in order to generate smaller molecules (e.g. glucose, fatty acids and amino acids to be used for fuels for energy use or other purposes). Water is thus essential and central to these metabolic processes. Therefore, without water, these metabolic processes would cease to exist, leaving us to muse about what processes would be in its place, such as gas absorption, dust collection, etc.

Water is also central to photosynthesis and respiration. Photosynthetic cells use the sun's energy to split off water's hydrogen from oxygen. Hydrogen is combined with CO_2 (absorbed from air or water) to form glucose and release oxygen. All living cells use such fuels and oxidise the hydrogen and carbon to capture the sun's energy and reform water and CO_2 in the process (cellular respiration).

Water is also central to acid-base neutrality and enzyme function. An acid, a hydrogen ion (H^+, that is, a proton) donor, can be neutralised by a base, a proton acceptor such as hydroxide ion (OH^-) to form water. Water is considered to be neutral, with a pH (the negative log of the hydrogen ion concentration) of 7. Acids have pH values less than 7 while bases have values greater than 7. Stomach acid (HCl) is useful to digestion. However, its corrosive effect on the esophagus during reflux can temporarily be neutralised by ingestion of a base such as aluminum hydroxide to produce the neutral molecules water and the salt aluminum chloride. Human biochemistry that involves enzymes usually performs optimally around a biologically neutral pH of 7.4.

For example a cell of Escherichia coli contains 70% of water, a human body 60-70%, plant body up to 90% and the body of an adult jellyfish is made up of 94–98% water.

Earth's waters are filled with life. The earliest life forms appeared in water; nearly all fish live exclusively in water, and there are many types of marine mammals, such as dolphins and whales that also live in the water. Some kinds of animals, such as amphibians, spend portions of their lives in water and portions on land. Plants such as kelp and algae grow in the water and are the basis for some underwater ecosystems. Plankton is generally the foundation of the ocean food chain.

Aquatic animals must obtain oxygen to survive, and they do so in various ways. Fish have gills instead of lungs, although some species of fish, such as the lungfish, have both. Marine mammals, such as dolphins, whales, otters, and seals need to surface periodically to breathe air. Smaller life forms are able to absorb oxygen through their skin.

EFFECTS ON HUMAN CIVILISATION

Civilisation has historically flourished around rivers and major waterways; Mesopotamia, the so-called cradle of civilisation, was situated between the major rivers Tigris and Euphrates; the ancient society of the Egyptians depended entirely upon the Nile.

Large metropolises like Rotterdam, London, Montreal, Paris, New York City, Shanghai, Tokyo, Chicago, and Hong Kong owe their success in part to their easy accessibility via water and the resultant expansion of trade. Islands with safe water ports, like Singapore, have flourished for the same reason. In places such as North Africa and the Middle East, where water is more scarce, access to clean drinking water was and is a major factor in human development.

Health and Pollution

Water fit for human consumption is called drinking water or potable water. Water that is not potable can be made potable by filtration or distillation (heating it until it becomes water vapour, and then capturing the vapour without any of the impurities it leaves behind), or by other methods (chemical or heat treatment that kills bacteria). Sometimes the term safe water is applied to potable water of a lower quality threshold (i.e., it is used effectively for nutrition in humans that have weak access to water cleaning processes, and does more good than harm).

Water that is not fit for drinking but is not harmful for humans when used for swimming or bathing is called by various names other than potable or drinking water, and is sometimes called safe water, or "safe for bathing". Chlorine is a skin and mucous membrane irritant that is used to make water safe for bathing or drinking. Its use is highly technical and is usually monitored by government regulations (typically 1 part per million (ppm) for drinking water, and 1-2 ppm of chlorine not yet reacted with impurities for bathing water).

This natural resource is becoming scarcer in certain places, and its availability is a major social and economic concern. Currently, about 1 billion people around the world routinely drink unhealthy water. Most countries accepted the goal of halving by 2015 the number of people worldwide who do not have access to safe water and sanitation during the 2003 G8 Evian summit. Even if this difficult goal is met, it will still leave more than an estimated half a billion people without access to safe drinking water and over 1 billion without access to adequate sanitation.

Poor water quality and bad sanitation are deadly; some 5 million deaths a year are caused by polluted drinking water. Water, however, is not a finite resource, but rather re-circulated as potable water in precipitation in quantities many degrees of magnitude higher than human consumption. Therefore, it is the relatively small quantity of water in reserve in the earth (about 1% of our drinking water supply, which is replenished in aquifers around every 1 to 10 years), that is a non-renewable resource, and it is, rather, the distribution of potable and irrigation water which is scarce, rather than the actual amount of it that exists on the earth.

Water-poor countries use importation of goods as the primary method of importing water (to leave enough for local human consumption), since the manufacturing process uses around 10 to 100 times products' masses in water. In the developing world, 90% of all wastewater still goes untreated into local rivers and streams. Some 50 countries, with roughly a third of the world's population, also suffer from medium or high water stress, and 17 of these extract more water annually than is recharged through their natural water cycles. The strain not only affects surface freshwater bodies like rivers and lakes, but it also degrades groundwater resources.

Human Uses

Agriculture

The most important use of water in agriculture is for an irrigation and irrigation is key component to produce enough food. Irrigation takes up to 90% of water withdrawn in some developing countries.

As a scientific standard

On 7 April 1795, the gram was defined in France to be equal to "the absolute weight of a volume of pure water equal to a cube of one hundredth of a meter, and to the temperature of the melting ice." For practical purposes though, a metallic reference standard was required, one thousand times more massive, the kilogram. Work was therefore commissioned to determine precisely how massive one liter of water was.

In spite of the fact that the decreed definition of the gram specified water at 0 °C—a highly stable *temperature* point—the scientists chose to redefine the standard and to perform their measurements at the most stable *density* point: the temperature at which water reaches maximum density, which was measured at the time as 4 °C. The Kelvin temperature scale of the SI system is based on the triple point of water. The scale is a more accurate development of the Celsius temperature scale, which is defined by the boiling point (100 °C) and melting point (0 °C) of water.

Natural water consists mainly of the isotopes hydrogen-1 and oxygen-16, but there is also small quantity of heavier hydrogen-2 (deuterium). The amount of deuterium oxides or heavy water is very small, but it still affects the properties of water. Water from rivers and lakes tends to contain less deuterium than seawater. Therefore, a standard water called Vienna Standard Mean Ocean Water is defined as the standard water.

For drinking

The human body is anywhere from 55% to 78% water depending on body size. To function properly, the body requires between one and seven liters of water per day to avoid dehydration; the precise amount depends on the level of activity, temperature, humidity, and other factors. Most of this is ingested through foods or beverages other than drinking straight water. It is not clear how much water intake is needed by healthy people, though most advocates agree that 6–7 glasses of water (approximately 2 litres) daily is the minimum to maintain proper hydration.

Medical literature favours a lower consumption, typically 1 liter of water for an average male, excluding extra requirements due to fluid loss from exercise or warm weather. For those who have healthy kidneys, it is rather difficult to drink too much water, but (especially in warm humid weather and while exercising) it is dangerous to drink too little. People can drink far more water than necessary while exercising, however, putting them at risk of water intoxication (hyperhydration), which can be fatal. The "fact" that a person should consume eight glasses of water per day cannot be traced back to a scientific source. There are other myths such as the effect of water on weight loss and constipation that have been dispelled.

An original recommendation for water intake in 1945 by the Food and Nutrition Board of the National Research Council read: "An ordinary standard for diverse persons is 1 milliliter for each calorie of food. Most of this quantity is contained in prepared foods." The latest dietary reference intake report by the United States National Research Council in general recommended: 2.7 liters of water total for women and 3.7 liters for men. Specifically, pregnant and breastfeeding women need additional fluids to stay hydrated.

According to the Institute of Medicine—who recommend that, on average, women consume 2.2 litres and men 3.0 litres—this is recommended to be 2.4 litres (approx. 9 cups) for pregnant women and 3 litres (approx. 12.5 cups) for breastfeeding women since an especially large amount of fluid is lost during nursing. Also noted is that normally, about 20 percent of water intake comes from food, while the rest comes from drinking water and beverages (caffeinated included). Water is excreted from the body in multiple forms; through urine and feces, through sweating, and by exhalation of water vapour in the breath. With physical exertion and heat exposure, water loss will increase and daily fluid needs may increase as well.

Humans require water that does not contain too many impurities. Common impurities include metal salts and/or harmful bacteria, such as Vibrio. Some solutes are acceptable and even desirable for taste enhancement and to provide needed electrolytes.

The single largest freshwater resource suitable for drinking is Lake Baikal in Siberia, which has a very low salt and calcium content and is very clean.

As a dissolving agent or solvent

Dissolving (or suspending) is used to wash everyday items such as the human body, clothes, floors, cars, food, and pets. Also, human wastes are carried by water in the sewage system. Its use as a cleaning solvent consumes most of water in industrialised countries.

Water can facilitate the chemical processing of wastewater. An aqueous environment can be favourable to the breakdown of pollutants, due to the ability to gain an homogenous solution that is pumpable and flexible to treat. Aerobic treatment can be used by applying oxygen or air to a solution reduce the reactivity of substances within it.

Water also facilitates biological processing of waste that have been dissolved within it. Microorganisms that live within water can access dissolved wastes and can feed upon them breaking them down into less polluting substances. Reedbeds and anaerobic digesters are both examples of biological systems that are particularly suited to the treatment of effluents.

Typically from both chemical and biological treatment of wastes, there is often a solid residue or cake that is left over from the treatment process. Depending upon its constituent parts, this 'cake' may be dried and spread on land as a fertiliser if it has beneficial properties, or alternatively disposed of in landfill or incinerated.

Water is the most abundant molecule in organisms. Fruits shrink when they are dried because they consist primarily of water.

As a heat transfer fluid

Water and steam are used as heat transfer fluids in diverse heat exchange systems, due to its availability and high heat capacity, both as a coolant and for heating. Cool water may even be naturally available from a lake or the sea. Condensing steam is a particularly efficient heating fluid because of the large heat of vaporisation. A disadvantage is that water and steam are somewhat corrosive. In almost all electric power plants, water is the coolant, which vaporises and drives steam turbines to drive generators.

In the nuclear industry, water can also be used as a neutron moderator. In a pressurised water reactor, water is both a coolant and a moderator. This provides a passive safety measure, as removing the water from the reactor also slows the nuclear reaction down.

Extinguishing fires

Water has a high heat of vaporisation and is relatively inert, which makes it a good fire extinguishing fluid. The evaporation of water carries heat away from the fire. However, water cannot be used to fight fires of electric equipment, because impure water is electrically conductive, or of oils and organic solvents, because they float on water and the explosive boiling of water tends to spread the burning liquid.

Decomposition of water may have played a role in the Chernobyl disaster. Initially, cooling of the incandescent reactor was attempted, but the result was an explosion, when the extreme heat caused water to flash into steam, thus leading to a steam explosion; it may also have decomposed water into hydrogen and oxygen, which subsequently exploded.

Chemical uses

Organic reactions are usually quenched with water or a water solution of a suitable acid, base or buffer. Water is generally effective in removing inorganic salts. In inorganic reactions, water is a common solvent. In organic reactions, it is usually not used as a reaction solvent, because it does not dissolve the reactants well and is amphoteric (acidic *and* basic) and nucleophilic. Nevertheless, these properties are sometimes desirable. Also, acceleration of Diels-Alder reactions by water has been observed. Supercritical water has recently been a topic of research. Oxygen-saturated supercritical water combusts organic pollutants efficiently.

Recreation

Humans use water for many recreational purposes, as well as for exercising and for sports. Some of these include swimming, waterskiing, boating, and diving. In addition, some sports, like ice hockey and ice skating, are played on ice. Lakesides, beaches and waterparks are popular places for people to go to relax and enjoy recreation. Many find the sound of flowing water to be calming, too. Some keep fish and other life in aquariums or ponds for show, fun, and companionship. Humans also use water for snow sports i.e. skiing or snowboarding, which requires the water to be frozen. People may also use water for play fighting such as with snowballs, water guns or water balloons. They may also make fountains and use water in their public or private decorations.

Water industry

The water industry provides drinking water and wastewater services to households and industry.

Water supply facilities includes for example water wells cisterns for rainwater harvesting, water supply network, water purification facilities, water tanks, water towers, water pipes including old aqueducts. Atmospheric water generator is in development.

Drinking water is often collected at springs, extracted from artificial borings in the ground, or wells. Building more wells in adequate places is thus a possible way to produce more water, assuming the aquifers can supply an adequate flow. Other water sources are rainwater and river or lake water. This surface water, however, must be purified for human consumption. This may involve removal of undissolved substances, dissolved substances and harmful microbes. Popular methods are filtering with sand which only removes undissolved material, while chlorination and boiling kill harmful microbes. Distillation does all three functions. More advanced techniques exist, such as reverse osmosis. Desalination of abundant ocean or seawater is a more expensive solution used in coastal arid climates.

The distribution of drinking water is done through municipal water systems or as bottled water. Governments in many countries have programs to distribute water to the needy at no charge. Others argue that the market mechanism and free enterprise are best to manage this rare resource and to finance the boring of wells or the construction of dams and reservoirs.

Reducing waste by using drinking water only for human consumption is another option. In some cities such as Hong Kong, sea water is extensively used for flushing toilets citywide in order to conserve fresh water resources.

Polluting water may be the biggest single misuse of water; to the extent that a pollutant limits other uses of the water, it becomes a waste of the resource, regardless of benefits to the polluter. Like other types of pollution, this does not enter standard accounting of market costs, being conceived as externalities for which the market cannot account. Thus other people pay the price of water pollution, while the private firms' profits are not redistributed to the local population victim of this pollution. Pharmaceuticals consumed by humans often end up in the waterways and can have detrimental effects on aquatic life if they bioaccumulate and if they are not biodegradable.

Wastewater facilities are sewers and wastewater treatment plants. Another way to remove pollution from surface runoff water is bioswale.

Industrial applications

Water is used in power generation. Hydroelectricity is electricity obtained from hydropower. Hydroelectric power comes from water driving a water turbine connected to a generator. Hydroelectricity is a low-cost, non-polluting, renewable energy source. The energy is supplied by the sun. Heat from the sun evaporates water, which condenses as rain in higher altitudes, from where it flows down.

Pressurised water is used in water blasting and water jet cutters. Also, very high pressure water guns are used for precise cutting. It works very well, is relatively safe, and is not harmful to the environment. It is also used in the cooling of machinery to prevent over-heating, or prevent saw blades from over-heating.

Water is also used in many industrial processes and machines, such as the steam turbine and heat exchanger, in addition to its use as a chemical solvent. Discharge of untreated water from industrial uses is pollution. Pollution includes discharged solutes (chemical pollution) and discharged coolant water (thermal pollution). Industry requires pure water for many applications and utilises a variety of purification techniques both in water supply and discharge.

Food processing

Water plays many critical roles within the field of food science. It is important for a food scientist to understand the roles that water plays within food processing to ensure the success of their products.

Solutes such as salts and sugars found in water affect the physical properties of water. The boiling and freezing points of water is affected by solutes. One mole of sucrose (sugar) raises the boiling point of water by 0.52 °C, and one mole of salt raises the boiling point by 1.04 °C while lowering the freezing point of water in a similar way. Solutes in water also affect water activity which affects many chemical reactions and the growth of microbes in food. Water activity can be described as a ratio of the vapour pressure of water in a solution to the vapour pressure of pure water. Solutes in water lower water activity. This is important to know because most bacterial growth ceases at low levels of water activity. Not only does microbial growth affect the safety of food but also the preservation and shelf life of food.

Water hardness is also a critical factor in food processing. It can dramatically affect the quality of a product as well as playing a role in sanitation. Water hardness is classified based on the amounts of removable calcium carbonate salt it contains per gallon. Water hardness is measured in grains; 0.064 g calcium carbonate is equivalent to one grain of hardness. Water is classified as soft if it contains 1 to 4 grains, medium if it contains 5 to 10 grains and hard if it contains 11 to 20 grains. The hardness of water may be altered or treated by using a chemical ion exchange system. The hardness of water also affects its pH balance which plays a critical role in food processing. For example, hard water prevents successful production of clear beverages. Water hardness also affects sanitation; with increasing hardness, there is a loss of effectiveness for its use as a sanitiser.

Boiling, steaming, and simmering are popular cooking methods that often require immersing food in water or its gaseous state, steam. While cooking water is used for dishwashing too.

WATER TREATMENT

Water treatment describes a process used to make water more acceptable for a desired end-use. These can include use as drinking water, industrial processes, medical and many other uses. The goal of all water treatment process is to remove existing contaminants in the water, improving it for subsequent use.

The goal may be to allow treated water to discharge into the without adverse ecological impact. These processes may be physical such as settling, chemical such as disinfection or coagulation, or biological such as lagooning, slow sand filtration or activated sludge.

There are a variety of waste treatment technologies. The long established method of waste treatment is simply landfill. Other technologies include composting and recycling. There are also a number of new and emerging waste treatment technologies which are able to treat a variety of waste streams.

There is a move away from treating waste as a disposal problem towards recognising waste contains valuable resources that can be recovered. Recovery of value from waste can either be in the form of recovery of resources or recovery of energy.

Water Purification

Water purification is the removal of contaminants from untreated water to produce drinking water that is pure enough for its intended use, most commonly human consumption. Substances that are removed during the process of drinking water treatment include bacteria, algae, viruses, fungi, minerals such as iron and sulphur, and man-made chemical pollutants.

Sewage Treatment

Sewage treatment is the process that removes the majority of the contaminants from wastewater or sewage and produces both a liquid effluent suitable for disposal to the natural environment and a sludge. To be effective, sewage must be conveyed to a treatment plant by appropriate pipes and infrastructure and the process itself must be subject to regulation and controls.

Some wastewaters require different and sometimes specialised treatment methods. At the simplest level, treatment of sewage and most wastewaters is carried out through separation of solids from liquids, usually by settlement. By progressively converting dissolved material into solids, usually a biological floc which is then settled out, an effluent stream of increasing purity is produced.

Irrigation

Irrigation is the artificial application of water to the soil usually for assisting in growing crops. In crop production it is mainly used in dry areas and in periods of rainfall shortfalls, but also to protect plants against frost.

Additionally irrigation helps to suppress weed growing in rice fields. In contrast, agriculture that relies only on direct rainfall is referred to as rain-fed farming. Irrigation is often studied together with drainage, which is the natural or artificial removal of surface and sub-surface water from a given area.

Irrigation is also a term used in the Medical/Dental fields and refers to flushing and washing out anything with water or another liquid.

CONCEPT OF WATER STRESS

The concept of water stress is relatively simple: According to the World Business Council for Sustainable Development, it applies to situations where there is not enough water for all uses, whether agricultural, industrial or domestic. Defining thresholds for stress in terms of available water per capita is more complex, however, entailing assumptions about water use and its efficiency. Nevertheless, it has been proposed that when annual per capita renewable freshwater availability is less than 1,700 cubic meters, countries begin to experience periodic or regular water stress. Below 1,000 cubic meters, water scarcity begins to hamper economic development and human health and well-being.

Population Growth

In 2000, the world population was 6.2 billion. The UN estimates that by 2050 there will be an additional 3 billion people with most of the growth in developing countries that already suffer water stress. Thus, water demand will increase unless there are corresponding increases in water conservation and recycling of this vital resource.

Increased Affluence

The rate of poverty alleviation is increasing especially within the two population giants of China and India. However, increasing affluence inevitably means more water consumption: from needing clean fresh water 24 hours a day, 7 days a week and basic sanitation service, to demanding water for gardens and car washing, to wanting jacuzzis or private swimming pools.

Expansion of Business Activity

Business activity ranging from industrialisation to services such as tourism and entertainment continues to expand rapidly. This expansion requires increased water services including both supply and sanitation, which can lead to more pressure on water resources and natural ecosystems.

Rapid Urbanisation

The trend towards urbanisation is accelerating. Small private wells and septic tanks that work well in low-density communities are not feasible within high-density urban areas. Urbanisation requires significant investment in water infrastructure in order to deliver water to individuals and to process the concentrations of wastewater – both from individuals and from business. These polluted and contaminated waters must be treated or they pose unacceptable public health risks. In 60% of European cities with more than 100,000 people, groundwater is being used at a faster rate than it can be replenished. Even if some water remains available, it costs more and more to capture it.

Climate Change

Climate change could have significant impacts on water resources around the world because of the close connections between the climate and hydrologic cycle. Rising temperatures will increase evaporation and lead to increases in precipitation, though there will be regional variations in rainfall. Overall, the global supply of freshwater will increase. Both droughts and floods may become more frequent in different regions at different times, and dramatic changes in snowfall and snowmelt are expected in mountainous areas. Higher temperatures will also affect water quality in ways that are not well understood. Possible impacts include increased eutrophication. Climate change could also mean an increase in demand for farm irrigation, garden sprinklers, and perhaps even swimming pools.

Depletion of Aquifers

Due to the expanding human population, competition for water is growing such that many of the worlds major aquifers are becoming depleted. This is due both for direct human consumption as well as agricultural irrigation by groundwater. Millions of small pumps of all sizes are currently extracting groundwater throughout the world. Irrigation in dry areas such as northern China and India is supplied by groundwater, and is being extracted at an unsustainable rate. Cities that have experienced aquifer drops between 10 to 50 meters include Mexico City, Bangkok, Manila, Beijing, Madras and Shanghai.

Pollution and Water Protection

Water pollution is one of the main concerns of the world today. The governments of many countries have striven to find solutions to reduce this problem. Many pollutants threaten water supplies, but the most widespread, especially in underdeveloped countries, is the discharge of raw sewage into natural waters; this method of sewage disposal is the most common method in underdeveloped countries, but also is prevalent in quasi-developed countries such as China, India and Iran.

Sewage, sludge, garbage, and even toxic pollutants are all dumped into the water. Even if sewage is treated, problems still arise. Treated sewage forms sludge, which is sent out into the sea and dumped. In addition to sewage, chemicals dumped by industries and governments are another major source of water pollution.

REFERENCES

Debenedetti, P. G., and Stanley, H. E.; "Supercooled and Glassy Water", *Physics Today* 56 (6), p. 40-46, 2003.

Franks, F (Ed), *Water, A comprehensive treatise,* Plenum Press, New York, 1972-1982.

Gleick PH and associates, *The World's Water: The Biennial Report on Freshwater Resources,* Island Press, Washington, D.C., 1998.

Marks, William E., "The Holy Order of Water: Healing Earth's Waters and Ourselves," Bell Pond Books (a div. of Steiner Books), Great Barrington, MA, November 2001.

2

Water Purification Technologies

Water purification may remove: particulate sand; suspended particles of organic material; parasites, Giardia; Cryptosporidium; bacteria; algae; virus; fungi; etc. Minerals calcium, silica, magnesium, etc. and toxic metals (lead, copper, chromium etc). Some purification may be elective in the purification process, including smell (hydrogen sulfide remediation), taste (mineral extraction), and appearance (iron incapsulation).

Governments usually dictate the standards for drinking water quality. These standards will require minimum / maximum set points of contaminants and the inclusion of control elements that produce drinking water. Quality standards in many countries require specific amounts of disinfectant (such as chlorine or ozone) in the water after it leaves the water treatment plant (WTP), to reduce the risk of re-contamination while the water is in the distribution system.

It is not possible to tell whether water is safe to drink just by looking at it. Simple procedures such as boiling or the use of a household activated carbon filter are not sufficient for treating all the possible contaminants that may be present in water from an unknown source. Even natural spring water—considered safe for all practical purposes in the 1800s—must now be tested before determining what kind of treatment, if any, is needed.

Chemical analysis, while expensive, is the only way to obtain the information necessary for deciding on method of purification.

SOURCES OF DRINKING WATER

Groundwater: The water emerging from some deep ground water may have fallen as rain many decades, hundreds, thousands or in some cases millions of years ago. Soil and rock layers naturally filter the ground water to a high degree of clarity before it is pumped to the treatment plant. Such water may emerge as springs, artesian springs, or may be extracted from boreholes or wells. Deep ground water is generally of very high bacteriological quality (i.e., pathogenic bacteria or the pathogenic protozoa are typically absent), but the water typically is rich in dissolved solids, especially carbonates and sulfates of calcium and magnesium.

Depending on the strata through which the water has flowed, other ions may also be present including chloride, and bicarbonate. There may be a requirement to reduce the iron or manganese content of this water to make it pleasant for drinking, cooking, and laundry use. Disinfection may also be required. Where groundwater recharge is practised, it is equivalent to lowland surface waters for treatment purposes.

Upland lakes and reservoirs: Typically located in the headwaters of river systems, upland reservoirs are usually sited above any human habitation and may be surrounded by a protective zone to restrict the opportunities for contamination. Bacteria and pathogen levels are usually low, but some bacteria, protozoa or algae will be present. Where uplands are forested or peaty, humic acids can colour the water. Many upland sources have low pH which require adjustment.

Rivers, canals and low land reservoirs: Low land surface waters will have a significant bacterial load and may also contain algae, suspended solids and a variety of dissolved constituents.

Atmospheric water generation is a new technology that can provide high quality drinking water by extracting water from the air by cooling the air and thus condensing water vapour. Rainwater harvesting or fog collection which collect water from the atmosphere can be used especially in areas with significant dry seasons and in areas which experience fog even when there is little rain.

Pre-treatment

Pumping and containment: The majority of water must be pumped from its source or directed into pipes or holding tanks. To avoid adding contaminants to the water, this physical infrastructure must be made from appropriate materials and constructed so that accidental contamination does not occur.

Screening: The first step in purifying surface water is to remove large debris such as sticks, leaves, trash and other large particles which may interfere with subsequent purification steps. Most deep groundwater does not need screening before other purification steps.

Storage: Water from rivers may also be stored in bankside reservoirs for periods between a few days and many months to allow natural biological purification to take place. This is especially important if treatment is by slow sand filters. Storage reservoirs also provide a buffer against short periods of drought or to allow water supply to be maintained during transitory pollution incidents in the source river.

Pre-conditioning: Many waters rich in hardness salts are treated with soda-ash (Sodium carbonate) to precipitate calcium carbonate out utilising the common ion effect.

Pre-chlorination: In many plants the incoming water was chlorinated to minimise the growth of fouling organisms on the pipe-work and tanks. Because of the potential adverse quality effects, this has largely been discontinued.

Widely varied techniques are available to remove the fine solids, micro-organisms and some dissolved inorganic and organic materials. The choice of method will depend on the quality of the water being treated, the cost of the treatment process and the quality standards expected of the processed water.

pH Adjustment

Distilled water has an average pH of 7 (neither alkaline nor acidic) and sea water has an average pH of 8.3 (slightly alkaline). If the water is acidic (lower than 7), lime or soda ash is added to raise the pH. Lime is the more common of the two additives because it is cheap, but it also adds to the resulting water hardness.

Making the water slightly alkaline ensures that coagulation and flocculation processes work effectively and also helps to minimise the risk of lead being dissolved from lead pipes and lead solder in pipe fittings. If the water is alkaline, acid (HCl) and carbon Dioxide (CO_2) are commonly added to lower pH. Having an alkaline water does not assure that deposits on the pipe i.e. lead or copper are not released into the water. Changing the electrolytic properties of the water are more indicative to release lead or copper into the water.

FLOCCULATION PROCESS

Flocculation is a process which clarifies the water. Clarifying means removing any turbidity or colour so that the water is clear and colourless. Clarification is done by causing a precipitate to form in the water which can be removed using simple physical methods. Initially the precipitate forms as very small particles but as the water is gently stirred, these particles stick together to form bigger particles—this process is sometimes called flocculation.

Many of the small particles that were originally present in the raw water absorb onto the surface of these small precipitate particles and so get incorporated into the larger particles that coagulation produces. In this way the coagulated precipitate takes most of the suspended matter out of the water and is then filtered off, generally by passing the mixture through a coarse sand filter or sometimes through a mixture of sand and granulated anthracite (high carbon and low valoatiles coal). Coagulants or flocculating agents that may be used include:

— *Iron (III) hydroxide.* This is formed by adding a solution of an iron (III) compound such as iron (III) chloride to pre-treated water with a pH of 7 or greater. Iron (III) hydroxide is extremely insoluble and forms even at a pH as low as 7. Commercial formulations of iron salts were traditionally marketed in the UK under the name *Cuprus*

Aluminium hydroxide is also widely used as the flocculating precipitate although there have been concerns about possible health impacts and mis-handling lead to a severe poisoning incident in 1988 at Camelford in south-west UK when the coagulant was introduced directly into the holding reservoir of final treated water.

Aluminium Hydroxychloride is an artificially produced polymer and is one of a class of synthetic polymers that are now widely used. These polymers have a high molecular weight and form very stable and readily removed flocs but tend to be more expensive in use compared to inorganic materials.

SEDIMENTATION

Water exiting the flocculation basin may enter the sedimentation basin, also called a clarifier or settling basin. It is a large tank with slow flow, allowing floc to settle to the bottom. The sedimentation basin is best located close to the flocculation basin so the transit between does not permit settlement or floc break up.

Sedimentation basins can be in the shape of a rectangle, where water flows from end to end, or circular where flow is from the centre outward. Sedimentation basin outflow is typically over a weir so only a thin top layer-furthest from the sediment-exits. The amount of floc that settles out of the water is dependent on the time the water spends in the basin and the depth of the basin. The retention time of the water must therefore be balanced against the cost of a larger basin. The minimum clarifier retention time is normally 4 hours.

A deep basin will allow more floc to settle out than a shallow basin. This is because large particles settle faster than smaller ones, so large particles bump into and integrate smaller particles as they settle. In effect, large particles sweep vertically though the basin and clean out smaller particles on their way to the bottom. As particles settle to the bottom of the basin a layer of sludge is formed on the floor of the tank. This layer of sludge must be removed and treated.

The amount of sludge that is generated is significant, often 3%-5% of the total volume of water that is treated. The cost of treating and disposing of the sludge can be a significant part of the operating cost of a water treatment plant. The tank may be equipped with mechanical cleaning devices that continually clean the bottom of the tank or the tank can be taken out of service when the bottom needs to be cleaned.

WATER FILTRATION

After separating most floc, the water is filtered as the final step to remove remaining suspended particles and unsettled floc. The most common type of filter is a rapid sand filter. Water moves vertically through sand which often has a layer of activated carbon or anthracite coal above the sand. The top layer removes organic compounds, which contribute to taste and odour.

The space between sand particles is larger than the smallest suspended particles, so simple filtration is not enough. Most particles pass through surface layers but are trapped in pore spaces or adhere to sand particles. Effective filtration extends into the depth of the filter. This property of the filter is key to its operation: if the top layer of sand were to block all the particles, the filter would quickly clog.

To clean the filter, water is passed quickly upward through the filter, opposite the normal direction (called *backflushing* or *backwashing*) to remove embedded particles. Prior to this, compressed air may be blown up through the bottom of the filter to break up the compacted filter media to aid the backwashing process; this is known as *air scouring*. This contaminated water can be disposed of, along with the sludge from the sedimentation basin, or it can be recycled by mixing with the raw water entering the plant.

Some water treatment plants employ pressure filters. These work on the same principle as rapid gravity filters, differing in that the filter medium is enclosed in a steel vessel and the water is forced through it under pressure.

Advantages

— Filters out much smaller particles than paper and sand filters can.

— Filters out virtually all particles larger than their specified pore sizes.

— They are quite thin and so liquids flow through them fairly rapidly.

— They are reasonably strong and so can withstand pressure differences across them of typically 2-5 atmospheres.

— They can be cleaned (back flushed) and reused.

Membrane filters are widely used for filtering both drinking water and sewage (for reuse). For drinking water, membrane filters can remove virtually all particles larger than 0.2 um—including Giardia and cryptosporidium. Membrane filters are an effective form of tertiary treatment when it is desired to reuse the water for industry, for limited domestic purposes, or before discharging the water into a river that is used by towns further downstream. They are widely used in industry, particularly for beverage preparation (including bottled water). However no filtration can remove substances that are actually dissolved in the water such as phosphorus, nitrates and heavy metal ions.

Slow sand Filters

Slow sand filters may be used where there is sufficient land and space as the water must be passed very slowly through the filters. These filters rely on biological treatment processes for their action rather than physical filtration. The filters are carefully constructed using graded layers of sand with the coarsest sand, along with some gravel, at the bottom and finest sand at the top. Drains at the base convey treated water away for disinfection. Filtration depends on the development of a thin biological layer, called the zoogleal layer or Schmutzdecke, on the surface of the filter.

An effective slow sand filter may remain in service for many weeks or even months if the pre-treatment is well designed and produces water with a very low available nutrient level which physical methods of treatment rarely achieve. Very low nutrient levels allow water to be safely sent through distribution system with very low disinfectant levels thereby reducing consumer irritation over offensive levels of chlorine and chlorine by-products. Slow sand filters are not backwashed; they are maintained by having the top layer of sand scraped off when flow is eventually obstructed by biological growth.

A specific 'large-scale' form of slow sand filter is the process of bank filtration, in which natural sediments in a riverbank are used to provide a first stage of contaminant filtration. While typically not sufficiently clean enough to be used directly for drinking water, the water gained from the associated extraction wells is much less problematic than river water taken directly from the major streams which bank filtration is often used at.

Lava Filters

Lava filters are similar to sand filters and may also only be used where there is sufficient land and space. Like sand filters, the filters rely on biological treatment processes for their action rather than physical filtration. Unlike slow sand filters however, they are constructed out of 2 layers of lava pebbles and a top layer of nutrient-free soil (only at the plants roots). On top, water-purifying plants (as Iris pseudacorus and Sparganium erectum) are placed. Usually, around 1/4 of the dimension of lavastone is required to purify the water and just like slow sand filters, a series of herringbone drains are placed (with lava filters these are placed at the bottom layer).

Ultrafiltration

Ultrafiltration membranes are a relatively new development; they use polymer film with chemically formed microscopic pores that can be used in place of granular media to filter water effectively without coagulants. The type of membrane media determines how much pressure is needed to drive the water through and what sizes of micro-organisms can be filtered out.

DISTILLATION

Distillation is probably the oldest method of water purification. Water is first heated to boiling. The water vapour rises to a condenser where cooling water lowers the temperature so the vapour is condensed, collected and stored. Most contaminants remain behind in the liquid phase vessel. However, organics with boiling points lower than 100 °C cannot be removed efficiently and can actually become concentrated in the product water. Another disadvantage is cost. Distillation requires large amounts of energy and water.

Table 1. Distillation

Advantages	*Disadvantages*
Removes a broad range of contaminants	Some contaminants can be carried into the condensate
Reusable	Requires careful maintenance to ensure purity
	Consumes large amounts of energy

ION EXCHANGE PROCESS

The ion exchange process percolates water through bead-like spherical resin materials (ion-exchange resins). Ions in the water are exchanged for other ions fixed to the beads. The two most common ion-exchange methods are softening and deionisation. Softening is used primarily as a pretreatment method to reduce water hardness prior to reverse osmosis (RO) processing. The softeners contain beads that exchange two sodium ions for every calcium or magnesium ion removed from the "softened" water.

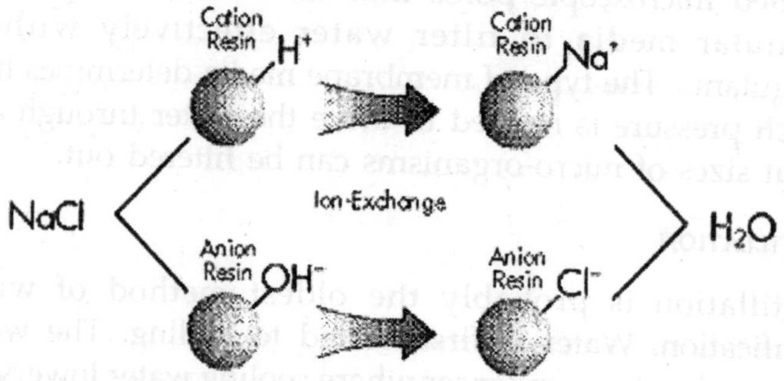

Figure 1

Deionisation (DI) beads exchange either hydrogen ions for cations or hydroxyl ions for anions. The cation exchange resins, made of styrene and divinylbenzene containing sulfonic acid groups, will exchange a hydrogen ion for any cations they encounter (e.g., Na^+, Ca^{++}, Al^{+++}). Similarly, the anion exchange resins, made of styrene and containing quaternary ammonium groups, will exchange a hydroxyl ion for any anions (e.g., Cl^-). The hydrogen ion from the cation exchanger unites with the hydroxyl ion of the anion exchanger to form pure water.

These resins may be packaged in separate bed exchangers with separate units for the cation and anion exchange beds. Or, they may be packed in mixed bed exchangers containing a mixture of both types of resins. In either case, the resin must be "regenerated" once it has exchanged all its hydrogen and/or hydroxyl ions for charged contaminants in the water. This regeneration reverses the purification process, replacing the contaminants bound to the DI resins with hydrogen and hydroxyl ions.

Table 2. Deionisation

Advantages	*Disadvantages*
Removes dissolved inorganics effectively.	Does not effectively remove particles, pyrogens or bacteria.
Regenerable (service deionisation).	DI beds can generate resin particles and culture bacteria.
Relatively inexpensive initial capital investment.	High operating costs over long-term.

Deionisation can be an important component of a total water purification system when used in combination with other methods discussed in this primer such as RO, filtration and carbon adsorption. DI systems effectively remove ions, but they do not effectively remove most organics or microorganisms. Microorganisms can attach to the resins, providing a culture media for rapid bacterial growth and subsequent pyrogen generation. The advantages and disadvantages of this technology are summarised above.

CARBON ADSORPTION

Organics can be cationic, anionic or nonionic. Ion-exchange resins remove some soluble organic acids and bases (anions and cations) from raw water, but some nonionic organics coat the resin. This process, known as resin "fouling," decreases the life of the resin and diminishes its capacity. To protect the ion-exchange resin, carbon filters can be placed upstream to remove nonionic organics.

The carbon adsorption process is controlled by the diameter of the pores in the carbon filter and by the diffusion rate of organic molecules through the pores. The rate of adsorption is a function of the molecular weight and the molecular size of the organics. Certain granular carbons effectively remove chloramines.

Carbon also removes free chlorine and protects other purification media in the system that may be sensitive to an oxidant such as chlorine. Carbon is usually used in combination with other treatment processes. The placement of carbon in relation to other components is an important consideration in the design of a water purification system.

Table 3. Carbon Adsorption

Advantages	*Disadvantages*
Removes dissolved organics and chlorine effectively.	Can generate carbon fines.
Long life (high capacity).	

MICROPOROUS FILTRATION

There are three types of microporous filtration: depth, screen and surface. Depth filters are matted fibers or materials compressed to form a matrix that retains particles by random adsorption or entrapment. Screen filters are inherently uniform structures which, like a sieve, retain all particles larger than the precisely controlled pore size on their surface. Surface filters are made from multiple layers of media. When fluid passes through the filter, particles larger than the spaces within the filter matrix are retained, accumulating primarily on the surface of the filter.

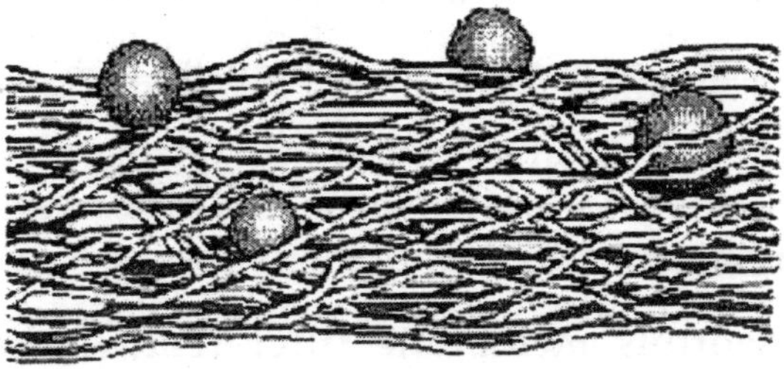

Figure 2

The distinction between filters is important because the three serve very different functions. Depth filters are usually used as prefilters because they are an economical way to remove 98% of suspended solids and protect elements downstream from fouling or clogging.

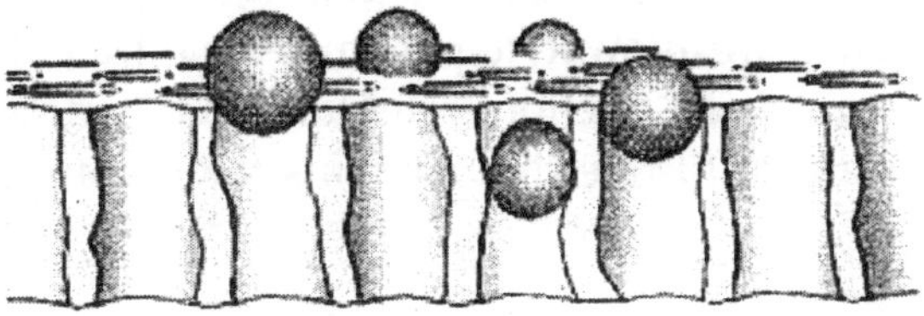

Figure 3

Surface filters remove 99.99% of suspended solids and may be used as either prefilters or clarifying filters. Microporous membrane (screen) filters are placed at the last possible point in a system to remove the last remaining traces of resin fragments, carbon fines, colloidal particles and microorganisms. For example, 0.22 μm Millipore membrane filters, which retain all bacteria, are routinely used to sterilise intravenous solutions, serums and antibiotics.

Table 4. Microporous Membrane Filtration

Advantages	*Disadvantages*
Absolute filters remove all particles and microorganisms greater than the pore size.	Will not remove dissolved inorganics, pyrogens or all colloidals.
Requires minimal maintenance	Potentially high expendable costs.
	Not regenerable.

ULTRAFILTRATION

A microporous membrane filter removes particles according to pore size. By contrast, an ultrafiltration (UF) membrane functions as a molecular sieve. It separates dissolved molecules on the basis of size by passing a solution through an infinitesimally fine filter.

The ultrafilter is a tough, thin, selectively permeable membrane that retains most macromolecules above a certain size including colloids, microorganisms and pyrogens. Smaller molecules, such as solvents and ionised contaminants, are allowed to pass into the filtrate. Thus, UF provides a retained fraction (retentate) that is rich in large molecules and a filtrate that contains few, if any, of these molecules.

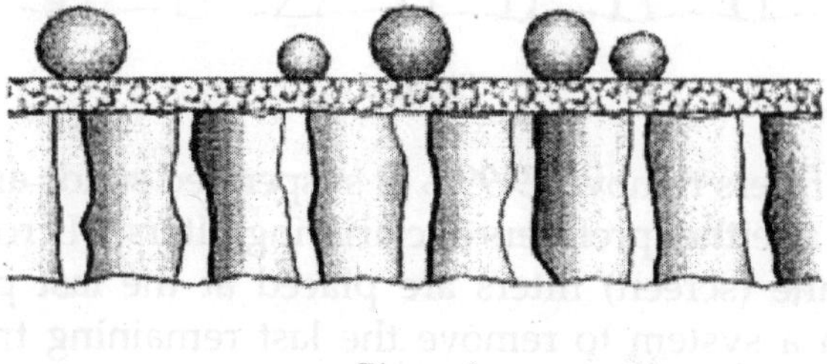

Figure 4

Ultrafilters are available in several selective ranges. In all cases, the membranes will retain most, but not necessarily all, molecules above their rated size.

Table 5. Ultrafiltration

Advantages	*Disadvantages*
Effectively removes most particles, pyrogens, microorganisms, and colloids above their rated size.	Will not remove dissolved inorganics.
Produces highest quality water for least amount of energy.	
Regenerable.	

REVERSE OSMOSIS (RO)

Reverse osmosis is the most economical method of removing 90% to 99% of all contaminants. The pore structure of RO membranes is much tighter than UF membranes. RO membranes are capable of rejecting practically all particles, bacteria and organics >300 daltons molecular weight (including pyrogens).

Natural osmosis occurs when solutions with two different concentrations are separated by a semi-permeable membrane. Osmotic pressure drives water through the membrane; the water dilutes the more concentrated solution; and the end result is an equilibrium.

In water purification systems, hydraulic pressure is applied to the concentrated solution to counteract the osmotic pressure. Pure water is driven from the concentrated solution and collected downstream of the membrane. Because RO membranes are very restrictive, they yield very slow flow rates. Storage tanks are required to produce an adequate volume in a reasonable amount of time.

RO also involves an ionic exclusion process. Only solvent is allowed to pass through the semi-permeable RO membrane, while virtually all ions and dissolved molecules are retained (including salts and sugars). The semi-permeable membrane rejects salts (ions) by a charge phenomena action: the greater the charge, the greater the rejection. Therefore, the membrane rejects nearly all (>99%) strongly ionised polyvalent ions but only 95% of the weakly ionised monovalent ions like sodium.

Different feed water may require different types of RO membranes. Membranes are manufactured from cellulose acetate or thin-film composites of polyamide on a polysulfone substrate. RO is the most economical and efficient method for purifying tap water if the system is properly designed for the feed water conditions and the intended use of the product water. RO is also the optimum pretreatment for reagent-grade water polishing systems.

Table 6. Reverse Osmosis

Advantages	*Disadvantages*
Effectively removes all types of contaminants to some extent (particles, pyrogens, microorganisms, colloids and dissolved inorganics). Requires minimal maintenance.	Limited flow rates.

ELECTRODEIONISATION

This new technology is a combination of electrodialysis and ion exchange, resulting in a process which effectively deionises water while the ion exchange resins are continuously regenerated by the electric current in the unit. This electrochemical regeneration replaces the chemical regeneration of conventional ion exchange systems.

The Elix™ module consists of a number of "cells" sandwiched between two electrodes. Each cell consists of a polypropylene frame onto which are bonded a cation-permeable membrane on one side, and an anion-permeable membrane on the other. The space in the centre of the cell, between the ion-selective membranes, is filed with a thin bed of ion exchange resins. The cells are separated from one another by a screen separator. The feed water entering the module is split into three parts. A small percentage flows over the electrodes, 65-75% of the feed passes through the resin beds in the cell, and the remainder passes along the screen separator between the cells.

Table 7. Electrodeionisation

Advantages	*Disadvantages*
Removes dissolved inorganics effectively	Needs pre-purification.
Non-polluting and safe:	
— No chemical regeneration	
— No chemical disposal	
— No resin disposal	
Inexpensive to operate	

The ion-exchange resins capture dissolved ions in the feed water at the top of the cell. Electric current applied across the module pulls those ions through the ion-selective membrane towards the electrodes. Cations are pulled through the cation-permeable membrane towards the cathode, and anions through the anion-selective membrane towards the anode.

These ions, however, are unable to travel all the way to their respective electrodes since they come to the adjacent ion-selective membrane which is of the opposite charge. This prevents further migrations of ions, which are then forced to concentrate in the space between the cells. This space is known as the "concentrate" channel, and the ions concentrated in this area are flushed out of the system to the drain.

The channel running through the resin bed in the centre of the cell is known as the "dilute" channel. As water passes down this channel, it is progressively deionised. At the lower end of the dilute channel, where water is free of ions, splitting of H_2O occurs in the electric field. This generates H^+ and OH^- which regenerate the ion exchange resins, effectively eliminating chemical regeneration.

ULTRAVIOLET (UV) RADIATION

Ultraviolet radiation has widely been used as a germicidal treatment for water. Mercury low pressure lamps generating 254 nm UV light are an effective means of sanitising water. The adsorption of UV light by the DNA and proteins in the microbial cell results in the inactivation of the microorganism.

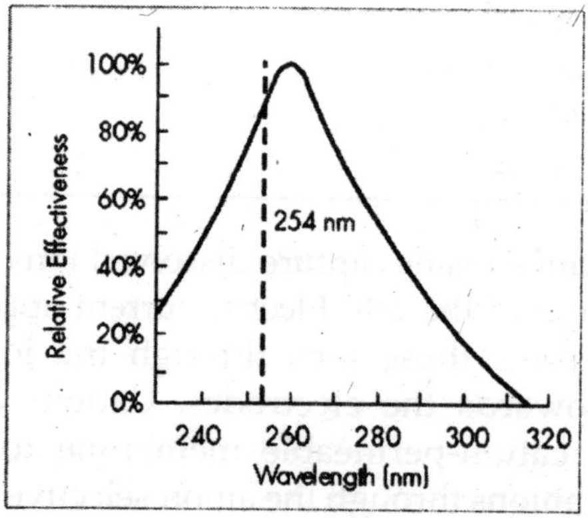

Figure 6

Recent advances in UV lamp technology have resulted in the production of special lamps which generate both 185 nm and 254 nm UV light. This combination of wavelengths is necessary for the photooxidation of organic compounds. With these special lamps, Total Organic Carbon (TOC) levels in high purity water can be reduced to 5 ppb.

Table 8. Ultraviolet Radiation

Advantages	*Disadvantages*
Effective sanitising treatment.	Decreases resistivity.
Oxidation of organic compounds (185 nm and 254 nm) to < 5 ppb TOC	Will not remove particles, colloids, or ions

PULLING IT ALL TOGETHER

Water Purification Systems

Because each purification technology removes a specific type of contaminant, none can be relied upon to remove all contaminants to the levels required for critical applications. A well-designed water purification system uses a combination of purification technologies to achieve final water quality.

Each of the purification technologies must be used in an appropriate sequence to optimise their particular removal capabilities. The schematic below shows a central laboratory water purification system designed to produce water for critical applications.

The first step is pretreatment equipment specifically designed to remove contaminants in the feed water. Pretreatment removes contaminants that may affect purification equipment located downstream, especially reverse osmosis (RO) systems. Examples of pretreatment are: carbon filters (or tanks) for chlorine removal, particulate filters for sediment/silt/particulate removal, and softening agents to remove minerals that cause "hard" water.

The next purification step is RO. RO removes 90 to 99% of all the contaminants found in water. It is the heart of any well designed water purification system because it effectively removes a broad range of contaminants.

However, the tight porosity of the RO membrane limits its flow rate. Therefore, a storage container is used to collect water from the system and distribute it to other points-of-use such as polishing systems.

Polishing systems purify pretreated water, such as RO water, by removing trace levels of any residual contaminants. Polishing elevates the quality of pretreated water to "Type I" or "ultrapure" water. A polishing system is designed to remove residual traces of impurities from water already pretreated by some other means (such as reverse osmosis or deionisation). Treating raw tap water using such a system would quickly exhaust its capacity and affect final quality.

A typical polishing system may consist of activated carbon, mixed-bed deionisation, organic scavenging mixtures and 0.22 µm final filtration. Systems can also be enhanced with ultrafiltration, ultraviolet oxidation or other features for use in specific applications. This combination of purification technologies, combined with proper pretreatment, will produce water that is virtually free of ionic, organic and microbial contamination.

MECHANICAL AND BIOLOGICAL TECHNIQUES

In addition to the many techniques used in large-scale water treatment, several small-scale, less (or non)-polluting techniques are also being used to treat polluted water. These techniques include techniques based on mechanical and biological processes. An overview:

— *mechanical systems:* sand filtration, lava filter systems and systems based on UV-radiation)

— biological systems:

— plant systems as constructed wetlands and treatment ponds (sometimes incorrectly called reedbeds) and living walls) and

— compact systems as activated sludge systems, biorotors, aerobic and anaerobic biofilters, submerged aerated filters, biorolls.

In order to purify the water aduquatly, several of these systems are usually combined to work as a whole. Combination of the systems is done in 2 to 3 stages, knowingly a primary and a secundary purification. Sometimes a tertairy purification is also added.

DISINFECTION

Disinfection is accomplished both by filtering out harmful microbes and also by adding disinfectant chemicals in the last step in purifying drinking water. Water is disinfected to kill any pathogens which pass through the filters. Possible pathogens include viruses, bacteria, including Escherichia coli, Campylobacter and Shigella, and protozoans, including G. lamblia and other Cryptosporidia.

In most developed countries, public water supplies are required to maintain a residual disinfecting agent throughout the distribution system, in which water may remain for days before reaching the consumer. Following the introduction of any chemical disinfecting agent, the water is usually held in temporary storage—often called a contact tank or clear well to allow the disinfecting action to complete.

Chlorination

The most common disinfection method is some form of chlorine or its compounds such as chloramine or chlorine dioxide. Chlorine is a strong oxidant that rapidly kills many harmful micro-organisms.

Because chlorine is a toxic gas, there is a danger of a release associated with its use. This problem is avoided by the use of sodium hypochlorite, which is a relatively inexpensive solution that releases free chlorine when dissolved in water. Chlorine solutions can be generated on site by electrolysing common salt solutions. A solid form, calcium hypochlorite exists that releases chlorine on contact with water. Handling the solid, however, requires greater routine human contact through opening bags and pouring than the use of gas cylinders or bleach which are more easily automated.

The generation of liquid sodium hypochlorite is both inexpensive and safer than the use of gas or solid chlorine. All forms of chlorine are widely used despite their respective drawbacks. One drawback is that chlorine from any source reacts with natural organic compounds in the water to form potentially harmful chemical by-products trihalomethanes (THMs) and haloacetic acids (HAAs), both of which are carcinogenic in large quantities and regulated by the United States Environmental Protection Agency (EPA).

The formation of THMs and haloacetic acids may be minimised by effective removal of as many organics from the water as possible prior to chlorine addition. Although chlorine is effective in killing bacteria, it has limited effectiveness against protosoans that form cysts in water. (Giardia lamblia and Cryptosporidium, both of which are pathogenic).

Chlorine dioxide is another faster-acting disinfectant. It is, however, relatively rarely used, because in some circumstances it may create excessive amounts of chlorite, which is a by-product regulated to low allowable levels in the USA. Chlorine dioxide is made in water and added/ used in water to avoid gas handling problems; chlorine dioxide gas accumulations may spontaneously detonate.

Chloramines are another chlorine-based disinfectant. Although chloramine is not as strong of an oxidant it does provide a longer lasting residual than free chlorine, and it not form THMs or haloacetic acids. It is possible to convert chlorine to chloramine by adding ammonia to the water after addition of chlorine: The chlorine and ammonia react to form chloramine. Water distribution systems disinfected with chloramines may experience nitrification, wherein ammonia is used a nutrient for bacterial growth, with nitrates being generated as a byproduct.

Ozone (O_3) is a relatively unstable molecule "free radical" of oxygen which readily gives up one atom of oxygen providing a powerful oxidising agent which is toxic to most water borne organisms. It is a very strong, broad spectrum disinfectant that is widely used in Europe. It is an effective method to inactivate harmful protozoans that form cysts. It also works well against almost all other pathogens.

Ozone is made by passing oxygen through ultraviolet light or a "cold" electrical discharge. To use ozone as a disinfectant, it must be created on site and added to the water by bubble contact. Some of the advantages of ozone include the production of relatively fewer dangerous by-products (in comparison to chlorination) and the lack of taste and odour produced by ozonation. Although fewer by-products are formed by ozonation, it has been discovered that the use of ozone produces a small amount of the suspected carcinogen Bromate, although little Bromine should be present in treated water.

Another one of the main disadvantages of ozone is that it leaves no disinfectant residual in the water. Ozone has been used in drinking water plants since 1906 where the first industrial ozonation plant was built in Nice, France. The U.S. Food and Drug Administration has accepted ozone as being safe; and it is applied as an anti-microbiological agent for the treatment, storage, and processing of foods.

UV radiation (light) is very effective at inactivating cysts, as long as the water has a low level of colour so the UV can pass through without being absorbed. The main disadvantage to the use of UV radiation is that, like ozone treatment, it leaves no residual disinfectant in the water. Because neither ozone nor UV radiation leaves a residual disinfectant in the water, it is sometimes necessary to add a residual disinfectant after they are used. This is often done through the addition of chloramines, discussed above as a primary disinfectant. When used in this manner, chloramines provide an effective residual disinfectant with very little of the negative aspects of chlorination.

Hydrogen peroxide is another disinfectant. It works similar to ozone, yet activators as formic acid are to be added to increase the working of this chemical substance. It also has the disadvantage that it is slow working; phytotoxic (in high dosage) and that it decreases the PH of the water it purifies.

Additional Treatment Options

Fluoridation : in many areas fluoride is added to water for the purpose of preventing tooth decay. This process is referred to as water fluoridation. Fluoride is usually added after the disinfection process. In the United States, fluoridation is usually accomplished by the addition of hexafluorosilicic acid, which decomposes in water, yielding fluoride ions.

Water conditioning: This is a method of reducing the effects of hard water. Hardness salts are deposited in water systems subject to heating because the decomposition of bicarbonate ions creates carbonate ions which crystalise out of the saturated solution of calcium or magnesium carbonate. Water with high concentrations of hardness salts can be treated with soda ash (sodium carbonate) which precipitates out the excess salts, through the Common-ion effect, producing calcium carbonate of very high purity.

The precipitated calcium carbonate is traditionally sold to the manufacturers of toothpaste. Several other methods of industrial and residential water treatment are claimed (without general scientific acceptance) to include the use of magnetic or/and electrical fields reducing the effects of hard water.

Plumbosolvency reduction: In areas with naturally acidic waters of low conductivity (i.e surface rainfall in upland mountains of igneous rocks), the water may be capable of dissolving lead from any lead pipes that it is carried in. The addition of small quantities of phosphate ion and increasing the pH slightly both assist in greatly reducing plumbo-solvency by creating insoluble lead salts on the inner surfaces of the pipes.

Radium Removal: Some groundwater sources contain radium, a radioactive chemical element. Typical sources include many groundwater sources north of the Illinois River in Illinois. Radium can be removed by ion exchange, or by water conditioning. The back flush or sludge that is produced is, however, a low-level radioactive waste.

Fluoride Removal: Although fluoride is added to water in many areas, some areas of the world have excessive levels of natural fluoride in the source water. Excessive levels can be toxic or cause undesirable cosmetic effects such as staining of teeth. One method of reducing fluoride levels is through treatment with activated alumina.

WATER PURIFICATION METHODS

Other popular methods for purifying water, especially for local private supplies are listed below. In some countries some of these methods are also used for large scale municipal supplies. Particularly important are distillation (de-salination of seawater) and reverse osmosis.

Boiling: Water is heated hot enough and long enough to inactivate or kill micro-organisms that normally live in water at room temperature. Near sea level, a vigorous rolling boil for at least one minute is sufficient. At high altitudes (greater than two kilometers or 5000 feet) three minutes is recommended. In areas where the water is "hard" (that is, containing significant dissolved calcium salts), boiling decomposes the bicarbonate ions, resulting in partial precipitation as calcium carbonate. This is the "fur" that builds up on kettle elements, etc., in hard water areas. With the exception of calcium, boiling does not remove solutes of higher boiling point than water and in fact increases their concentration (due to some water being lost as vapour). Boiling does not leave a residual disinfectant in the water. Therefore, water that has been boiled and then stored for any length of time may have acquired new pathogens.

Granular Activated Carbon filtering: GAC, a form of activated carbon with a high surface area, absorbs many compounds including many toxic compounds. Water passing through activated carbon is commonly used in municipal regions with organic contamination, taste or ordors. Many household water filters and fish tanks use activated carbon filters to further purify the water. Household filters for drinking water sometimes contain silver to release silver ions which have an anti-bacterial effect.

Distillation involves boiling the water to produce water vapour. The vapour contacts a cool surface where it condenses as a liquid. Because the solutes are not normally vaporised, they remain in the boiling solution. Even distillation does not completely purify water, because of contaminants with similar boiling points and droplets of unvaporised liquid carried with the steam. However, 99.9% pure water can be obtained by distillation. Distillation does not confer any residual disinfectant and the distillation apparatus may be the ideal place to harbour Legionnaires' disease.

Reverse osmosis: Mechanical pressure is applied to an impure solution to force pure water through a semi-permeable membrane. Reverse osmosis is theoretically the most thorough method of large scale water purification available, although perfect semi-permeable membranes are difficult to create. Unless membranes are well-maintained, algae and other life forms can colonise the membranes.

Ion exchange: Most common ion exchange systems use a zeolite resin bed to replace unwanted Ca^{2+} and Mg^{2+} ions with benign (soap friendly) Na^{+} or K^{+} ions. This is the common water softener.

Electrodeionisation: Water is passed between a positive electrode and a negative electrode. Ion selective membranes allow the positive ions to separate from the water toward the negative electrode and the negative ions toward the positive electrode. High purity deionised water results. The water is usually passed through a reverse osmosis unit first to remove non-ionic organic contaminants.

Direct contact membrane distillation (DCMD). Applicable to desalination. Heated seawater is passed along the surface of a hydrophobic polymer membrane. Evaporated water passes from the hot side through pores in the membrane into a stream of cold pure water on the other side. The difference in vapour pressure between the hot and cold side helps to push water molecules through.

Portable Water Purification

Portable techniques for purifying water are used for hiking, camping etc. or for use in rural areas or emergency situations. Common techniques include boiling, disinfection with tablets or ultra-filtration using a small hand pump. In addition to this "portable water purification" may also include residential, commercial or industrial water treatment systems that are regenerated off site (by others). These systems are most commonly used for deionisation in industrial applications or softening for residential applications. Also known as Service Deionisation (SDI), industrial companies use these portable systems for a variety of reasons.

Most commonly, the user has a small application and can not justify the expense of having regenerant chemicals such as sulfuric or hydrochloric acid and sodium hydroxide on site for safety and or manpower reasons. There may be waste water discharge limitations that make owning and operating an in-house deioniser (also known as a demineraliser) too expensive in comparison to the outsourcing of regeneration.

In residential applications a water treatment company delivers portable softener exchange tanks to a household. These softener tanks treat all or some of the water entering the household and remove hardness minerals. They are exchanged on a pre-determined schedule so the household has softened water on a on a continual basis. Although not as cost effective as owning a water softener, portable water softeners require no maintenance by the home owner; furthermore there is no discharge of brine to septic or municipal waste treatment systems.

Safety and Controversies

There has been controversy over the fluoridation of water since the 1930s, when the benefits to dental health were first claimed. There is evidence that excessive concentrations of fluorides can make teeth, and bones, more brittle and easily damaged/broken. Fluoridation is banned in many countries. Water supplies have sometimes been the subject of concerns over terrorism or actual terrorist threats. Accidents have also been known to happen. In April, 2007, the water supply of Spencer, Massachusetts became contaminated with excess sodium hydroxide (lye) when its treatment equipment malfunctioned.

REFERENCES

Beychok, M.R., "Performance of surface-aerated basins", *Chemical Engineering Progress Symposium Series* 67 (107): 322-339, 1971.

Massoud Tajrishy and Ahmad Abrishamchi, *Integrated Approach to Water and Wastewater Management for Tehran, Iran, Water Conservation, Reuse, and Recycling: Proceedings of the Iranian-American Workshop*, National Academies Press, 2005.

3

Sewage Treatment

Sewage is created by residences, institutions, hospitals and commercial and industrial establishments. It can be treated close to where it is created, or collected and transported via a network of pipes and pump stations to a municipal treatment plant. Sewage collection and treatment is typically subject to local, state and federal regulations and standards. Industrial sources of wastewater often require specialised treatment processes.

The sewage treatment involves three stages, called *primary, secondary* and *tertiary treatment*. First, the solids are separated from the wastewater stream. Then dissolved biological matter is progressively converted into a solid mass by using indigenous, water-borne microorganisms. Finally, the biological solids are neutralised then disposed of or re-used, and the treated water may be disinfected chemically or physically. The final effluent can be discharged into a stream, river, bay, lagoon or wetland, or it can be used for the irrigation of a golf course, green way or park. If it is sufficiently clean, it can also be used for groundwater recharge.

SEWAGE SYSTEMS

Raw influent (sewage) includes household waste liquid from toilets, baths, showers, kitchens, sinks, and so forth that is disposed of via sewers. In many areas, sewage also includes liquid waste from industry and commerce. The draining of household waste into greywater and blackwater is becoming more common in the developed world, with greywater being permitted to be used for watering plants or recycled for flushing toilets. A lot of sewage also includes some surface water from roofs or hard-standing areas.

Municipal wastewater therefore includes residential, commercial, and industrial liquid waste discharges, and may include stormwater runoff. Sewage systems capable of handling stormwater are known as combined systems or combined sewers. Such systems are usually avoided since they complicate and thereby reduce the efficiency of sewage treatment plants owing to their seasonality. The variability in flow also leads to often larger than necessary, and subsequently more expensive, treatment facilities.

In addition, heavy storms that contribute more flows than the treatment plant can handle may overwhelm the sewage treatment system, causing a spill or overflow. It is preferable to have a separate storm drain system for stormwater in areas that are developed with sewer systems. As rainfall runs over the surface of roofs and the ground, it may pick up various contaminants including soil particles and other sediment, heavy metals, organic compounds, animal waste, and oil and grease.

Some jurisdictions require stormwater to receive some level of treatment before being discharged directly into waterways. Examples of treatment processes used for stormwater include sedimentation basins, wetlands, buried concrete vaults with various kinds of filters, and vortex separators. The site where the raw wastewater is processed before it is discharged back to the environment is called a wastewater treatment plant (WWTP). The order and types of mechanical, chemical and biological systems that comprise the wastewater treatment plant are typically the same for most developed countries:

- Mechanical treatment
  - Influx (Influent)
  - Removal of large objects
  - Removal of sand and grit
  - Pre-precipitation
- Biological treatment
  - Oxidation bed (oxidising bed) or aeration system
  - Post precipitation
- Chemical treatment (this step is usually combined with settling and other processes to remove solids, such as filtration. The combination is referred to in the U.S. as physical-chemical treatment.

TREATMENT STAGES

Primary Treatment

Primary treatment removes the materials that can be easily collected from the raw wastewater and disposed of. The typical materials that are removed during primary treatment include fats, oils, and greases, sand, gravels and rocks, larger settleable solids and floating materials (such as rags and flushed feminine hygiene products). This step is done entirely with machinery.

Removal of large objects from influent sewage

In primary treatment, the influent sewage water is strained to remove all large objects that are deposited in the sewer system, such as rags, sticks, tampons, cans, fruit, etc. This is most commonly done with a manual or automated mechanically raked screen. The raking action of a mechanical bar screen is typically paced according to the accumulation on the bar screens and/or flow rate. The bar screen is used because large solids can damage or clog the equipment used later in the sewage treatment plant. The solids are collected in a dumpster and later disposed in a landfill.

Sand and grit removal

Primary treatment also typically includes a sand or grit channel or chamber where the velocity of the incoming wastewater is carefully controlled to allow sand grit and stones to settle, while keeping the majority of the suspended organic material in the water column. This equipment is called a detritor or sand catcher.

Sand, grit, and stones need to be removed early in the process to avoid damage to pumps and other equipment in the remaining treatment stages. Sometimes there is a sand washer (grit classifier) followed by a conveyor that transports the sand to a container for disposal. The contents from the sand catcher may be fed into the incinerator in a sludge processing plant, but in many cases, the sand and grit is sent to a landfill.

Sedimentation

Many plants have a sedimentation stage where the sewage is allowed to pass slowly through large tanks, commonly called "primary clarifiers" or "primary sedimentation tanks". The tanks are large enough that sludge can settle and floating material such as grease and oils can rise to the surface and be skimmed off.

The main purpose of the primary clarification stage is to produce both a generally homogeneous liquid capable of being treated biologically and a sludge that can be separately treated or processed. Primary settling tanks are usually equipped with mechanically driven scrapers that continually drive the collected sludge towards a hopper in the base of the tank from where it can be pumped to further sludge treatment stages.

Secondary Treatment

Secondary treatment is designed to substantially degrade the biological content of the sewage such as are derived from human waste, food waste, soaps and detergent. The majority of municipal and industrial plants treat the settled sewage liquor using aerobic biological processes. For this to be effective, the biota require both oxygen and a substrate on which to live. There are number of ways in which this is done. In all these methods, the bacteria and protozoa consume biodegradable soluble organic contaminants (e.g. sugars, fats, organic short-chain carbon molecules, etc.) and bind much of the less soluble fractions into floc.

Secondary treatment systems are classified as fixed film or suspended growth. Fixed-film treatment process including trickling filter and rotating biological contactors where the biomass grows on media and the sewage passes over its surface. In suspended growth systems—such as activated sludge—the biomass is well mixed with the sewage and can be operated in a smaller space than fixed-film systems that treat the same amount of water. However, fixed-film systems are more able to cope with drastic changes in the amount of biological material and can provide higher removal rates for organic material and suspended solids than suspended growth systems.

Roughing filters are intended to treat particularly strong or variable organic loads, typically industrial, to allow them to then be treated by conventional secondary treatment processes. Characteristics include typically tall, circular filters filled with open synthetic filter media to which wastewater is applied at a relatively high rate. They are designed to allow high hydraulic loading and a high flow-through of air. On larger installations, air is forced through the media using blowers. The resultant wastewater is usually within the normal range for conventional treatment processes.

Activated sludge plants

In general, activated sludge plants encompass a variety of mechanisms and processes that use dissolved oxygen to promote the growth of biological floc that substantially removes organic material. The process traps particulate material and can, under ideal conditions, convert ammonia to nitrite and nitrate and ultimately to nitrogen gas.

Surface-aerated basins

Most biological oxidation processes for treating industrial wastewaters have in common the use of oxygen (or air) and microbial action. Surface-aerated basins achieve 80 to 90% removal of BOD with retention times of 1 to 10 days. The basins may range in depth from 1.5 to 5.0 metres and use motor-driven aerators floating on the surface of the wastewater. In an aerated basin system, the aerators provide two functions: they transfer air into the basins required by the biological oxidation reactions, and they provide the mixing required for dispersing the air and for contacting the reactants (that is, oxygen, wastewater and microbes).

Typically, the floating surface aerators are rated to deliver the amount of air equivalent to 1.8 to 2.7 kg O_2/kW·h. However, they do not provide as good mixing as is normally achieved in activated sludge systems and therefore aerated basins do not achieve the same performance level as activated sludge units. Biological oxidation processes are sensitive to temperature and, between 0 °C and 40 °C, the rate of biological reactions increase with temperature. Most surface aerated vessels operate at between 4 °C and 32 °C.

Fluidised bed reactors

The carbon absorption following biological treatment was particularly effective in reducing both the BOD and COD to low levels. A fluidised bed reactor is a combination of the most common stirred tank packed bed, continuous flow reactors. It is very important to chemical engineering because of its excellent heat and mass transfer characteristics. In a fluidised bed reactor, the substrate is passed upward through the immobilised enzyme bed at a high velocity to lift the particles. However the velocity must not be so high that the enzymes are swept away from the reactor entirely. This causes low mixing; these type of reactors are highly suitable for the exothermic reactions. It is most often applied in immobilised enzyme catalysis.

Filter beds

In older plants and plants receiving more variable loads, trickling filter beds are used where the settled sewage liquor is spread onto the surface of a deep bed made up of coke (carbonised coal), limestone chips or specially fabricated plastic media. Such media must have high surface areas to support the biofilms that form. The liquor is distributed through perforated rotating arms radiating from a central pivot. The distributed liquor trickles through this bed and is collected in drains at the base. These drains also provide a source of air which percolates up through the bed, keeping it aerobic.

Biological films of bacteria, protozoa and fungi form on the media's surfaces and eat or otherwise reduce the organic content. This biofilm is grazed by insect larvae and worms which help maintain an optimal thickness. Overloading of beds increases the thickness of the film leading to clogging of the filter media and ponding on the surface.

Biological aerated filters (BAF)

Biological Aerated (or Anoxic) Filter (BAF) or Biofilters combine filtration with biological carbon reduction, nitrification or denitrification. BAF usually includes a reactor filled with a filter media. The media is either in suspension or supported by a gravel layer at the foot of the filter. The dual purpose of this media is to support highly active biomass that is attached to it and to filter suspended solids. Carbon reduction and ammonia conversion occurs in aerobic mode and sometime achieved in a single reactor while nitrate conversion occurs in anoxic mode. BAF is operated either in upflow or downflow configuration depending on design specified by manufacturer.

Membrane bioreactors (MBR)

Membrane bioreactors (MBR) combines activated sludge treatment with a membrane liquid-solid separation process. The membrane component uses low pressure microfiltration or ultra filtration membranes and eliminates the need for clarification and tertiary filtration. The membranes are typically immersed in the aeration tank (however, some applications utilise a separate membrane tank). One of the key benefits of a membrane bioreactor system is that it effectively overcomes the limitations associated with poor settling of sludge in conventional activated sludge (CAS) processes.

The technology permits bioreactor operation with considerably higher mixed liquor suspended solids (MLSS) concentration than CAS systems, which are limited by sludge settling. The process is typically operated at MLSS in the range of 8,000–12,000 mg/L, while CAS are operated in the range of 2,000–3,000 mg/L. The elevated biomass concentration in the membrane bioreactor process allows for very effective removal of both soluble and particulate biodegradable materials at higher loading rates. Thus increased Sludge Retention Times (SRTs)—usually exceeding 15 days—ensure complete nitrification even in extremely cold weather.

The cost of building and operating a MBR is usually higher than conventional wastewater treatment, however, as the technology has become increasingly popular and has gained wider acceptance throughout the industry, the life-cycle costs have been steadily decreasing. As well, in developed urban areas where the footprint of the treatment plant is considered a limiting factor MBR facilities can be considered a desirable option.

Secondary sedimentation

The final step in the secondary treatment stage is to settle out the biological floc or filter material and produce sewage water containing very low levels of organic material and suspended matter.

Rotating biological contactors

Schematic diagram of a typical rotating biological contactor (RBC). The treated effluent clarifier/settler is not included in the diagram.

Rotating biological contactors (RBCs) are mechanical secondary treatment systems, which are robust and capable of withstanding surges in organic load. RBCs were first installed in Germany in 1960 and have since been developed and refined into a reliable operating unit. The rotating disks support the growth of bacteria and micro-organisms present in the sewage, which breakdown and stabilise organic pollutants.

To be successful, micro-organisms need both oxygen to live and food to grow. Oxygen is obtained from the atmosphere as the disks rotate. As the micro-organisms grow, they build up on the media until they are sloughed off due to shear forces provided by the rotating discs in the sewage. Effluent from the RBC is then passed through final clarifiers where the micro-organisms in suspension settle as a sludge. The sludge is withdrawn from the clarifier for further treatment.

Tertiary Treatment

Tertiary treatment provides a final stage to raise the effluent quality before it is discharged to the receiving environment (sea, river, lake, ground, etc.). More than one tertiary treatment process may be used at any treatment plant. If disinfection is practiced, it is always the final process. It is also called "effluent polishing".

Filtration

Sand filtration removes much of the residual suspended matter. Filtration over activated carbon removes residual toxins.

Lagooning

Lagooning provides settlement and further biological improvement through storage in large man-made ponds or lagoons. These lagoons are highly aerobic and colonisation by native macrophytes, especially reeds, is often encouraged. Small filter feeding invertebrates such as Daphnia and species of Rotifera greatly assist in treatment by removing fine particulates.

Constructed wetlands

Constructed wetlands include engineered reedbeds and a range of similar methodologies, all of which provide a high degree of aerobic biological improvement and can often be used instead of secondary treatment for small communities, also see phytoremediation. One example is a small reedbed used to clean the drainage from the elephants' enclosure at Chester Zoo in England.

Nutrient removal

Wastewater may contain high levels of the nutrients nitrogen and phosphorus. Excessive release to the environment can lead to a build up of nutrients, called eutrophication, which can in turn encourage the overgrowth of weeds, algae, and cyanobacteria (blue-green algae). This may cause an algal bloom, a rapid growth in the population of algae.

The algae numbers are unsustainable and eventually most of them die. The decomposition of the algae by bacteria uses up so much of oxygen in the water that most or all of the animals die, which creates more organic matter for the bacteria to decompose. In addition to causing deoxygenation, some algal species produce toxins that contaminate drinking water supplies. Different treatment processes are required to remove nitrogen and phosphorus.

Nitrogen removal

The removal of nitrogen is effected through the biological oxidation of nitrogen from ammonia (nitrification) to nitrate, followed by denitrification, the reduction of nitrate to nitrogen gas. Nitrogen gas is released to the atmosphere and thus removed from the water.

Nitrification itself is a two-step aerobic process, each step facilitated by a different type of bacteria. The oxidation of ammonia (NH_3) to nitrite (NO_2^-) is most often facilitated by *Nitrosomonas* spp. (nitroso referring to the formation of a nitroso functional group). Nitrite oxidation to nitrate (NO_3^-), though traditionally believed to be facilitated by *Nitrobacter* spp. (nitro referring the formation of a nitro functional group), is now known to be facilitated in the environment almost exclusively by *Nitrospira* spp.

Denitrification requires anoxic conditions to encourage the appropriate biological communities to form. It is facilitated by a wide diversity of bacteria. Sand filters, lagooning and reed beds can all be used to reduce nitrogen, but the activated sludge process (if designed well) can do the job the most easily. Since denitrification is the reduction of nitrate to dinitrogen gas, an electron donor is needed. This can be, depending on the wastewater, organic matter (from faeces), sulfide, or an added donor like methanol. Sometimes the conversion of toxic ammonia to nitrate alone is referred to as tertiary treatment.

Phosphorus removal

Phosphorus can be removed biologically in a process called enhanced biological phosphorus removal. In this process, specific bacteria, called polyphosphate accumulating organisms, are selectively enriched and accumulate large quantities of phosphorus within their cells (up to 20% of their mass). When the biomass enriched in these bacteria is separated from the treated water, these biosolids have a high fertiliser value.

Phosphorus removal can also be achieved by chemical precipitation, usually with salts of iron (e.g. ferric chloride) or aluminum (e.g. alum). The resulting chemical sludge is difficult to handle and the added chemicals can be expensive. Despite this, chemical phosphorus removal requires significantly smaller equipment footprint than biological removal, is easier to operate and can be more reliable in areas that have wastewater compositions that make biological phosphorus removal difficult.

Disinfection

The purpose of disinfection in the treatment of wastewater is to substantially reduce the number of microorganisms in the water to be discharged back into the environment. The effectiveness of disinfection depends on the quality of the water being treated (e.g., cloudiness, pH, etc.), the type of disinfection being used, the disinfectant dosage (concentration and time), and other environmental variables. Cloudy water will be treated less successfully since solid matter can shield organisms, especially from ultraviolet light or if contact times are low. Generally, short contact times, low doses and high flows all militate against effective disinfection. Common methods of disinfection include ozone, chlorine, or ultraviolet light. Chloramine, which is used for drinking water, is not used in wastewater treatment because of its persistence.

Chlorination remains the most common form of wastewater disinfection in North America due to its low cost and long-term history of effectiveness. One disadvantage is that chlorination of residual organic material can generate chlorinated-organic compounds that may be carcinogenic or harmful to the environment. Residual chlorine or chloramines may also be capable of chlorinating organic material in the natural aquatic environment. Further, because residual chlorine is toxic to aquatic species, the treated effluent must also be chemically dechlorinated, adding to the complexity and cost of treatment.

Ultraviolet (UV) light can be used instead of chlorine, iodine, or other chemicals. Because no chemicals are used, the treated water has no adverse effect on organisms that later consume it, as may be the case with other methods. UV radiation causes damage to the genetic structure of bacteria, viruses, and other pathogens, making them incapable of reproduction. The key disadvantages of UV disinfection are the need for frequent lamp maintenance and replacement and the need for a highly treated effluent to ensure that the target microorganisms are not shielded from the UV radiation (i.e., any solids present in the treated effluent may protect microorganisms from the UV light).

In the United Kingdom, light is becoming the most common means of disinfection because of the concerns about the impacts of chlorine in chlorinating residual organics in the wastewater and in chlorinating organics in the receiving water. Edmonton, Alberta, Canada also uses UV light for its water treatment. Ozone O_3 is generated by passing oxygen O_2 through a high voltage potential resulting in a third oxygen atom becoming attached and forming O_3. Ozone is very unstable and reactive and oxidises most organic material it comes in contact with, thereby destroying many pathogenic microorganisms.

Ozone is considered to be safer than chlorine because, unlike chlorine which has to be stored on site (highly poisonous in the event of an accidental release), ozone is generated onsite as needed. Ozonation also produces fewer disinfection by-products than chlorination. A disadvantage of ozone disinfection is the high cost of the ozone generation equipment and the requirements for special operators.

PACKAGE PLANTS AND BATCH REACTORS

In order to use less space, treat difficult waste, deal with intermittent flow or achieve higher environmental standards, a number of designs of hybrid treatment plants have been produced. Such plants often combine all or at least two stages of the three main treatment stages into one combined stage. In the UK, where a large number of sewage treatment plants serve small populations, package plants are a viable alternative to building discrete structures for each process stage. One type of system that combines secondary treatment and settlement is the sequencing batch reactor (SBR). Typically, activated sludge is mixed with raw incoming sewage and mixed and aerated.

The resultant mixture is then allowed to settle producing a high quality effluent. The settled sludge is run off and re-aerated before a proportion is returned to the head of the works. SBR plants are now being deployed in many parts of the world including North Liberty, Iowa, and Llanasa, North Wales. The disadvantage of such processes is that precise control of timing, mixing and aeration is required. This precision is usually achieved by computer controls linked to many sensors in the plant. Such a complex, fragile system is unsuited to places where such controls may be unreliable, or poorly maintained, or where the power supply may be intermittent.

Package plants may be referred to as *high charged* or *low charged*. This refers to the way the biological load is processed. In high charged systems, the biological stage is presented with a high organic load and the combined floc and organic material is then oxygenated for a few hours before being charged again with a new load. In the low charged system the biological stage contains a low organic load and is combined with floculate for a relatively long time.

SLUDGE TREATMENT AND DISPOSAL

The sludges accumulated in a wastewater treatment process must be treated and disposed of in a safe and effective manner. The purpose of digestion is to reduce the amount of organic matter and the number of disease-causing microorganisms present in the solids. The most common treatment options include anaerobic digestion, aerobic digestion, and composting. The choice of a wastewater solid treatment method depends on the amount of solids generated and other site-specific conditions. However, in general, composting is most often applied to smaller-scale applications followed by aerobic digestion and then lastly anaerobic digestion for the larger-scale municipal applications.

Anaerobic Digestion

Anaerobic digestion is a bacterial process that is carried out in the absence of oxygen. The process can either be thermophilic digestion, in which sludge is fermented in tanks at a temperature of 55°C, or mesophilic, at a temperature of around 36°C. Though allowing shorter retention time, thermophilic digestion is more expensive in terms of energy consumption for heating the sludge. One major feature of anaerobic digestion is the production of biogas, which can be used in generators for electricity production and/or in boilers for heating purposes.

Aerobic Digestion

Aerobic digestion is a bacterial process occurring in the presence of oxygen. Under aerobic conditions, bacteria rapidly consume organic matter and convert it into carbon dioxide. The operating costs are characteristically much greater for aerobic digestion because of the energy costs needed to add oxygen to the process.

Composting

Composting is also an aerobic process that involves mixing the sludge with sources of carbon such as sawdust, straw or wood chips. In the presence of oxygen, bacteria digest both the wastewater solids and the added carbon source and, in doing so, produce a large amount of heat.

Thermal Depolymerisation

Thermal depolymerisation uses hydrous pyrolysis to convert reduced complex organics to oil.

Sludge disposal

When a liquid sludge is produced, further treatment may be required to make it suitable for final disposal. Typically, sludges are thickened (dewatered) to reduce the volumes transported off-site for disposal. There is no process which completely eliminates the need to dispose of biosolids. There is, however, an additional step some cities are taking to superheat the wastewater sludge and convert it into small pelletised granules that are high in nitrogen and other organic materials. This product is then sold to local farmers and turf farms as a soil amendment or fertiliser, reducing the amount of space required to dispose of sludge in landfills.

Treatment in the receiving environment

Many processes in a wastewater treatment plant are

designed to mimic the natural treatment processes that occur in the environment, whether that environment is a natural water body or the ground. If not overloaded, bacteria in the environment will consume organic contaminants, although this will reduce the levels of oxygen in the water and may significantly change the overall ecology of the receiving water. Native bacterial populations feed on the organic contaminants, and the numbers of disease-causing microorganisms are reduced by natural environmental conditions such as predation exposure to ultraviolet radiation, for example.

Consequently, in cases where the receiving environment provides a high level of dilution, a high degree of wastewater treatment may not be required. However, recent evidence has demonstrated that very low levels of certain contaminants in wastewater, including hormones and synthetic materials such as phthalates that mimic hormones in their action, can have an unpredictable adverse impact on the natural biota and potentially on humans if the water is re-used for drinking water. In the US and EU, uncontrolled discharges of wastewater to the environment are not permitted under law, and strict water quality requirements are to be met. A significant threat in the coming decades will be the increasing uncontrolled discharges of wastewater within rapidly developing countries.

Sewage treatment in developing countries

There are few reliable figures on the share of the wastewater collected in sewers that is being treated in the world. In many developing countries the bulk of domestic and industrial wastewater is discharged without any treatment or after primary treatment only. In Latin America about 15% of collected wastewater passes through treatment plants. In Venezuela, a below average country in South America with respect to wastewater treatment, 97 percent of the country's sewage is discharged raw into the environment.

In a relatively developed Middle Eastern country such as Iran, Tehran's majority of population has totally untreated sewage injected to the city's groundwater. Most of sub-Saharan Africa is without wastewater treatment.

Water utilities in developing countries are chronically underfunded because of low water tariffs, the inexistence of sanitation tariffs in many cases, low billing efficiency and poor operational efficiency. In addition, wastewater treatment typically is the process within the utility that receives the least attention, partly because enforcement of environmental standards is poor. As a result of all these factors, operation and maintenance of many wastewater treatment plants is poor. This is evidenced by the frequent breakdown of equipment, shutdown of electrically operated equipment due to power outages or to reduce costs, and sedimentation due to lack of sludge removal.

Developing countries as diverse as Egypt, Algeria, China or Colombia have invested substantial sums in wastewater treatment without achieving a significant impact in terms of environmental improvement. Even if wastewater treatment plants are properly operating, it can be argued that the environmental impact is limited in cases where the assimilative capacity of the receiving waters is high, as it is often the case.

REFERENCES

Maude Barlow, Tony Clarke., *Blue Gold: The Fight to Stop the Corporate Theft of the World's Water*, 2003.

Reisner, Marc., *Cadillac Desert: The American West and Its Disappearing Water*, 1993.

Vandana Shiva., *Water Wars: Privatization, Pollution, and Profit*, 2002.

Worster, Donald., *Rivers of Empire: Water, Aridity, and the Growth of the American West*, 1992.

4

Water Management in Crop Production

Rainfall is the cheapest source of natural water-supply for crop plants. In India, however, rainfall is notoriously capricious, causing floods and droughts alternately. Its frequency distribution and amount are not in accordance with the needs of the crops. Artificial water-supply through irrigation on one occasion, and removal of excess water through drainage on another occasion, therefore, become imperative, if the crops are to be raised successfully.

Water management in India, thus, comprises irrigation or drainage or both, depending considerably on the environmental conditions, soil, crops, and climate. It is a situation-oriented entity. Water effects the performance of crops not only directly but also indirectly by influencing the availability of other nutrients, the timing of cultural operations, etc. Water and other production inputs interact with one another. In proper combinations, the crop yields can be boosted manifold under irrigated agriculture.

Water is a costly input when canals supply it. The constructing of dams and reservoirs, the conveying of water from storage points to the fields, the operating and the maintaining of canal systems involve huge expense. The misuse of water leads to the problems of water-logging, salt-imbalance, etc., thus rendering agricultural lands unproductive. Hence a proper appreciation of the relationship among soils, crops, climate and water resources for maximum crop production.

WATER RESOURCES

Taking the total geographical area of the country at 328 million hectares and the average annual rainfall at about 112 cm, the total annual precipitation in the country is estimated at about 3,700,000 million cubic meters. The south-westerly monsoon contributes over 80 per cent to the total precipitation in the country.

The Central Water and Power Commission, New Delhi, has estimated that of the total annual precipitation amounting to 800,000 million cubic meters seeps into the ground, about 1,700,000 million cubic meters flows into the rivers and the remaining amount of about 1,200,000 million cubic meters evaporates back into the atmosphere. The water, flowing on the surface and that seeping into the ground, forms the two major sources of water for irrigating crops.

Surface-Water Resources

A large number of rivers of various potentials and discharges are spread all over the country. The rivers in the north, which originate from the Himalayas, are snow-fed and, thus, have less seasonal fluctuations in their flow than the rivers in the other parts of the country. The flow in the rivers of the central and southern parts of the country depends entirely on the monsoon.

The rivers flow to the full during the rainy season (July to September) and their flow dwindles with the approach of the summer. The surface-flowing water needs to be trapped in ponds, tanks, lakes or artificial reservoirs when it is available in abundance so that it can be fruitfully used for irrigation during the rainless period to meet the water needs of crops. Of the annual surface flow of 1,700,000 million cubic meters, only about 666,000 million cubic meters can be utilised for the purpose of irrigation owing to the physiographical limitations.

Ground-Water Sources

Substantial supplies are also available from ground-water sources. Of the 800,000 million cubic meters of rain-water that seeps into the ground annually, about 430,000 million cubic meters of it is absorbed by the surface layers of the earth's crust and, thus, can be utilised directly by the vegetation in the process of evapo-transpiration and growth. The remaining 370,000 million cubic meters of rain-water percolates deep into the porous strata of the earth's crust, representing the gross annual enrichment of the underground water. This ground-water is tapped by digging or drilling shallow or deep wells and is lifted by using mechanical devices for irrigating the crops.

A precise quantitative inventory regarding the ground-water reserves is not available. Organisation such as the Geographical Survey of India, the Central Ground-Water Board and the State Tube-Wells and the Ground-Water Boards are engaged in this task. It has been estimated by the Central Ground-Water Board that the total ground water reserves are on the order of 55,000,000 million cubic meters out of which 425,740 million cubic meters have been assessed as the annual recharge from rain and canal seepage. The Task Force on Ground-Water Reserves of the Planning Commission has also endorsed these estimates.

All recharge to the ground-water is not available for withdrawal, since part of it is lost as sub-surface flow. After accounting from these losses, the gross available ground-water recharge is about 269,960 million cubic meters per annum. A part of this recharge is in the saline regions of the country and is unsuitable for use in agriculture owing to its poor quality. The net recharge available for ground-water development in India, therefore, is of the magnitude of about 267,500 million cubic meters per annum. The Working Group of the Planning Commission Task Force Ground-Water Reserves estimated that the usable ground-water potential would be only 75 to 80 per cent of the net ground-water recharge available and recommended a figure of 203,600 million cubic meters per annum as the long-term potential for ground-water development in India.

IRRIGATION DEVELOPMENT

Irrigation has been practiced in India since time immemorial. Wells and tanks, as known today's as 'minor irrigation works', were constructed by several kings of ancient times for public welfare. Significant development irrigation, however, can be stated to have commenced from 1850 onwards, with large-scale or major irrigation projects. These activities have been further accelerated after Independence in 1947. Today, India has about 34 million hectares of land under irrigation and this amounts to about one-fifth of the total cultivated area.

With utilisation water resources, both from the surface flows and from the ground-water recharge, the Second Irrigation Commission has calculated that the ultimate area that can be irrigated is of the order of about 82 million hectares in the country. Till the end of the Fourth Plan, about 45 per cent of the surface utilisable flow and about 20 per cent of the usable ground-water was proposed to be utilised to irrigate about 34 million hectares.

The programs for modifying weather, the desalinisation of sea-water and the National Water Grid, if enforced, will further increase the potential for irrigation manifold. Surface water for irrigation is obtained from flowing rivers, and from tanks, ponds, lakes or artificial reservoirs. The flows of rivers are directly diverted into canals, or high dams are built across the river to form first large canals for irrigation. The future development of irrigation aims at impounding the surplus flows of rivers by constructing dams for use during the dry periods.

Water from all these sources is conveyed to the field through lined or unlined canals, distributaries and minors through the final structure called outlet. All this conveyance system up to the outlet is built by the Irrigation Departments. From the outlet, water flows into small water-courses which are constructed, owned and managed by a group of farmers. During its conveyance, there are considerable losses through seepage, percolation and evaporation.

From a typical water-distributary system, the losses in the main canal vary from 10 to 15 per cent, and in the water-courses from 15 to 30 per cent. Thus the total losses from the source till the water reaches the farmers' fields may amount to 40-60 per cent. Ground-water is tapped by digging shallow and large diameter percolation wells or drilling deep tube-wells and lifting it to the surface. Shallow wells derive their water-supply from the surrounding area through seepage, percolation, high-water table, etc. the deep wells depend for their water on aquifers, which may have their source at some distance.

River valleys, canal-irrigated areas, low-lying places, natural vegetation and trees growing luxuriantly are indications of the-presence of ground-water resources. The rate at which the water can be pumped out from a well depends upon recharging rate which, in turn, depends upon the permeability of the surrounding area in the case of shallow wells and on the thickness and the magnitude of aquifers.

Utilisation of water resources: The scientific utilisation of water resources for crop production involves the consideration of the suitability of land and water for irrigation and then planning of crops and water-management practices commensurate with them. Water-management practices include irrigation and drainage. Irrigation comprises three fundamentals: how much to drain, how best to drain and how rapidly to drain under a given situation of soil, water, and crops.

SOIL-WATER SYSTEM

Available Water for Crop Plants

Soil is a heterogeneous mass and consists of three phases, viz. The solid phase, the liquid phase and the gaseous phase. Mineral matter, consisting of sand, silt and clay and organic matter, forms the solid phase which serves as a framework (matrix) with numerous pores of irregular shapes and different sizes holding air and water in various proportions. Soil is a porous medium, and serves as a water reservoir or bank. Water is deposited in this bank as rain or irrigation, and plants withdraw it during their growth. Water is retained by a soil particle in the form of a thin film around it, and in the numerous small pores of the soil matrix with forces, such as surface tension capillarity, cohesion and adhesion.

The salts present in soil water further add to these forces by way of osmotic pressure. Plants, therefore, need to exert at least an equal amount of force for extracting water from the soil mass for their growth. Immediately after rain or irrigation, water infiltrates into the soil and continues to move in the soil mass to deeper layers because of the gravitational force.

The downward movement of water practically ceases after a certain time. The water retained in the soil under this situation is termed 'field capacity' which forms the upper limit of the available soil moisture for crop plant. In other words, any further addition of water will not be retained by the soil, but will be lost through deep percolation beyond the roots of a crop, thus making it unavailable for the growth of its plants.

After the wetting of the soil, as evaporation and transpiration continue, the soil water goes on diminishing till a point is reached when plants are unable to extract it. The moisture content at this stage is termed 'permanent wilting-point' and this sets the lower limit of the availability of soil water.

In other words, any moisture below this point will not support plant growth. The range of soil water between the field capacity and the permanent wilting-point is termed 'available soil water for crop growth'. The values of the available water-holding capacity of different major soil groups are shown in Table 1. The available soil water-holding capacity increases mainly with the fineness of texture and the content of organic matter.

Table 1. Available water-holding capacity of different soil types

Soil texture	*Field capacity (FC)*	*Permanent wilting-point % (PWP)*	*Bulk density g/cc (BD)*	*Available water (mm) per metre depth of soil profile d=FC-PWP/ 100 x BD x soil depth*
Sandy	5 to 10	2 to 6	1.5 to 1.8	50 to 100
Sandy loam	10 to 18	4 to 10	1.4 to 1.6	90 to 160
Loam	18 to 25	8 to 14	1.3 to 1.5	140 to 220
Clay loam	24 to 32	11 to 16	1.3 to 1.4	170 to 250
Clay	32 to 40	15 to 22	1.2 to 1.4	200 to 280

Availability of soil water

Three classical hypotheses have been put forth for the relative availability of soil water in the available range.

i) Water availability and, consequently, the crop growth is equal and uniform over the entire range from the field capacity to the permanent wilting-point. This holds good generally for perennial species, such as orchard and tree crops whose dense root mass permeates the soil matrix thoroughly.

ii) Water availability and crop growth proceed uniformly from the field capacity to a certain critical point beyond which crop growth decreases rapidly till the permanent wilting-point is reached. This view holds good for most of the seasonal field crops maturing up to the seed stage.

iii) The availability of water and the rate of crop growth decrease gradually as the soil water content decreases from the field capacity to the permanent wilting-point. This holds good generally for most of the forage crops and those grown vegetatively.

The effects of the crop and climatic factors on the availability of soil water can also be significant. The responses of crops to soil water are ultimately reflected through the plant-water status, and the atmospheric aridity acts as the primary driving force for the absorption of moisture from the soil. A given crop loses water at different rayes under different conditions of evaporative demands of the climate.

Ideal Soil for Irrigation

For most crop plants, except rice, the ideal soil for irrigation is that which is deep, without any water-table, has high water-holding capacity, infiltration rate and permeability, and low salt content. The loams and clay loams are generally good soils for irrigation, since the run-off, the number of irrigations necessary and the investment for drainage are low as compared with those in the case of other soils. Any soil can be put under irrigated agriculture permanently but only with due care and caution which, on certain soils, may be beyond economic limits.

The heavy soils often need surface as well as subsoil drainage. The light sandy soils involve a high application of costly inputs besides a considerable wastage of water. In the case of rice, the soils with low percolation rates are ideal for economising on water. Land submergence has been found to be a beneficial practice for obtaining high rice yields. But during this submergence, huge quantities of water, as high as 60-70 per cent, are lost through deep percolation. In India, nearly half the water resources are diverted to rice alone. The proper selection of land for growing rice is, therefore, vital for utilising the water resources efficiently and economically.

Plant and Water Absorption

A plant extracts water from the soil with the hairs on the absorbing surface of their elongating roots, through osmosis and diffusion. Inside the plant, water moves along the concentration gradient across the root tissues to the xylem vessels and from there upwards to the stem and the leaves. On the surface of a leaf, there are numerous openings, stomata, through which water is lost in the form of vapour on account of the incident heat energy. The accumulated water vapour around the leaf tissues is further carried away by the wind into the atmosphere.

The process of loss of water from the plants through the stomata and plant surface is known as transpiration. Huge amounts of water, practically all that is absorbed, is lost in this way by the plants continuously throughout their life. The quantity of water utilised in metabolic activities by a plant is insignificantly small as compared with the amount lost through transpiration. The primary cause of transpiration is the solar energy available for the vaporisation of water from the leaf surface. Water is pulled out as a continuous film from the root hairs to the leaf surface because of the sink strength of the atmosphere.

Stomata are photosensitive cells and are open only during the day in most of the plant species. They close at night and hence there is practically no nocturnal transpiration. The stomata regulate transpiration to some extent by closing their apertures when the soil is dry and there is shortage of moisture in the atmosphere. The exchange of gases for photosynthesis also takes place through the stomata and, hence, the closure of stomata means no photosynthesis, i.e. no growth and accumulation of dry matter. Plant nutrients also move primarily through water in the plants and are translocated to different plant parts along with the transpiration stream. The closure of the stomata, therefore, also reduces the nutrient uptake.

Quality of Water

Criteria and their limits: In irrigated agriculture, the quality of water used for irrigation should receive adequate attention. Irrigation water, regardless of its source, always contains some soluble salts in it. Apart from the total concentration of the dissolved salts, the concentration of some of the individual salts, and especially of those which are most harmful to crops, is important in determining the suitability of water for irrigation.

The constituents usually determined by analysing irrigation water are the electrical conductivity for the total dissolved salts, soluble sodium percentage, sodium absorption ratio, the boron content, pH, cations, such as calcium, magnesium, sodium, potassium, and anions, such as carbonates, bicarbonates, sulphates, chlorides and nitrates. The salinity and sodium-hazard classes are shown in Table 2. The accepted limits for boron and residual carbonates are given in table 3.

Quality of water from different sources: Water from the rivers which flows over salt-affected areas or in the deltaic regions has a greater concentration of salts sometimes as high as 7,500 ppm or even more. The quality of tank or lake water depends mainly on the soil salinity in the watershed areas and the aridity of the place.

The quality of water from ground-water resources, i.e. from shallow or deep wells, is generally poor under the situation of :

i) high aridity - arid and semi-arid regions, with less than 45 cm of rainfall;

ii) high water-table and water-logged conditions ; and

iii) the vicinity of sea-water, as in the coastal region.

Ground-water in other areas is generally good. In saline regions, where the water from shallow aquifers is poor, the water from deeper strata may be good. Such water may be tapped by digging deep-tube wells.

Table 2. Salinity and sodium-hazard classes of irrigation water

Salinity class	*Electrical conductivity micromhos/cmm at 25°C*
C_1 Low	Less than 250
C_2 Moderate	250 to 750
C_3 Medium to high	750 to 2,250
C_4 High	2,250
C_5 Very high	5,000 to 20,000
Sodium class	*SAR*
S_1 Low	Less than 10
S_2 Moderate	10 to 15
S_3 High	18 to 26
S_4 Very high	More than 26

Table 3. Permissable limits of boron and residual carbonates

Boron class	*Sensitive crops*	*Semi-tolerant*	*Tolerant crops crops*	*Residual carbonate*	*Content meq/litre class*
1.	<0.33	<0.67	<1.0	Safe	<1.25
2.	0.33 to 0.67	0.67 to 1.32	1.0 to 2.0	Marginal	1.25 to 2.5
3.	0.67 to 1.0	1.33 to 2.0	2.0 to 3.0	Unsuitable	2.5
4.	1.0 to 1.25	2.0 to 2.50	3.0 to 3.75	—	—

The continual use of poor-quality water for irrigation water can be improved by adding certain substances, such as sulphuric acid or sulpher to reduce the pH to the desired level. Gypsum, which is available commercially, can be applied to the soil or mixed with water where poor-quality water, containing a high concentration of sodium salts, is used for irrigation. The permeability of soils can be improved by incorporating organic matter into them to facilitate the leaching down of the salts beyond the root zone.

CROP PRODUCTION AND WATER SALINITY

Choice of crops and varieties. A crop or a variety which is more tolerant to salt should be selected for growing under the cropping pattern adopted. Crops have been broadly grouped as follows, according to their salt-tolerance limits.

Tolerant species (over 5,000 ppm salts): Barley, sugar-beet, date-palm, rape, kale, cotton, Rhodes grass, sesbania.

Semi-tolerant species (2,500 to 5,000 ppm salts): Rice, sorghum, maize, barley, sunflower, lucerne, berseem or guar, safflower, onion, spinach, lettuce, carrot, cluster-been, wheat, pearl-millet or bajra, grasses.

Sensitive species (below 2,500 ppm salts): Peas, cabbage, vetch, beans, gram, peaches, grapes, orange, grapefruit, potato, tomato.

The limits of relative tolerance of crops to boron are as under :

— *Tolerant species (2 to 4 ppm):* Date-palm, sugar-cotton, lucerne, onion, turnip, cabbage, carrot and lettuce.

— *Semi-tolerant species (1 to 2 ppm):* Wheat, maize, barley, cotton, sunflower, potato, tomato, peas, beans and sweet-potato.

— *Sensitive species (0.3 to 1.0 ppm):* Apple, apricot, grape, orange, grapefruit, plum, pear, cherry and walnut.

Emergence from the soil, the seedling phase and flowering are three very critical stages in the growth of most of the field crops in the presence of salinity. Frequent irrigation and in amounts in excess of the storage capacity of a soil are essential for maintaining low moisture stress and to leach down the salts. Row crops should be grown near the furrow bottom where salt concentration is low. The plant population should also be enhanced by increasing the seed-rate and by reducing the spacing to compensate for the poor performance of individual plants.

Water, if only saline, can be used by diluting it with good quality water. If it is a solid soil, there will be a permeability problem. It is necessary to replace the sodium by calcium by applying suitable ammendments and organic matter for allowing more water to penetrate the soil, so that salts can be easily leached down beyond the root-zone. Adequate arrangements for drainage should be made to lower the water-table and to maintain a satisfactory salt balance in the root-zone. Fertiliser releasing nitrogen slowly with a calcium base may be preferred and applied in split doses by placement.

Planning water management: The water-management system on a farm includes farm inlet and outlet structures, conveyance lines, the laying out of irrigation and drainage ditches, the erecting of water-control structures, such as water gates, diversion boxes and drop structures, and the grading or the levelling of the land. It may also include the construction of storage tanks, the installation of pumps, etc.

Table 4. The proper assessment of availability of water, soils and crop factors

Water availability	*Soil*	*Crop*
Quantity per day	Texture	Type of root system
Rotation or turn period	Structure	Life-span
System and method	Depth up to	Consumptive water

of irrigation	the water-table	needs in relation to climate
Water quality class	Infiltration and permeability	Critical periods with respect to moisture
	Depth of a href=soils1.html>soil	Market value
	Slope of land	Yield response in relation to water-supply

After a proper assessment of these factors and topographic survey, a farm is divided into a convenient number of blocks and plots. Each block should have an independent irrigation and drainage channel and an approach road. There should be the minimum wastage of land under the field bunds and channels. The water flow in the channel should be non-silting and non-erosive.

Water can be conveyed from the storage point or the farm inlet to the point of use through earthen or lined water-courses or through metallic, canvas, cement or plastic pipes, hoses or tubes. Several lining materials of varying cost and durability are now available. These are mud, cement, bitumen, asphalt, bricks, plastic membranes, sodium compounds, etc. Conveyance losses are extremely high and lining is essential for using water economically. The water-courses should be at a higher elevation than the field level.

Land drainage: The provision of drains is a must on every irrigated farm for sustained farming. Damage caused by excess water and the consequent development of the salt problem is enormous and unnoticeable in early stages. Except rice, almost all crops receive a severe setback during their growth when excess water stagnates even for a period as short as 3 to 4 days. In heavy-rainfall areas, there should be adequate arrangements for surface drainage for the speedy disposal of water. In high-water-table areas, there should be subsoil drainage.

Spacing between the drains depends upon the permeability characteristics of the soil, excess water-tolerance capacity of the crop grown, the amount of water to be disposed of and its speed. The drains may be open or closed. The open drains are easy to construct and maintain, but they occupy considerable land. The closed drains are costly, but save land and, hence, are desirable where the land is scarce.

The closed drains may be of mole type or tile type. A mole drain is an unlined channel in the subsoil made with a tractor-drawn mole plough. This type of drainage is suitable only on heavy clay soils. These drains last for two to three years. The installation of the mole system is cheaper than the tile system.

Tile drainage is rather an expensive system but it lasts several decades. The investment varies with the area to be drained, the depth of installation, the quantity of water to be handled and the type of soil, and the cost may range from Rs 1,500 to 3,000 per hectare. A tile is made of clay or concrete and it is about 30 to 50 cm in length and 7 to 12 cm in diameter. These are placed end to end, with a gap of two to three mm between them. Excess water enters the system through this space and is conveyed along the gradient. The preperation of land for irrigation also includes grading or levelling. Several implements, such as scrapers, floats and land planes are now available for moving the earth and smoothening the fields. These can be operated with animal power or tractors.

PRINCIPLES IN IRRIGATION

The term water requirement includes the amount of water for meeting the needs of evaporation, transpiration and metabolic activities (all together known as consumptive use), losses during the application of irrigation water and water needed for special operations, such as land preparation, transplanting, the leaching of excess water, etc.

Water requirement = Consumptive use of water + Application losses + Special needs

$$WR = A + B + C$$

The entity A, viz. the consumptive use depends upon the factors of weather, the type of crop canopy, the soil moisture status and the stage of the crop. The entity B, i.e. the application losses, depends upon the type of the irrigation system, the soil texture and structure, and management practices. The entity C, viz., special needs, depends upon such factors as the soil moisture status of the soil as well as water, and the nature of the crop species grown.

Recent researches have revealed that meteorological factors play a very prominent role in deciding the magnitude of the consumptive use of water which forms a major part of the value of water requirements of crop plants. Evapo-transpiration involves the conversion of soil moisture into water vapour and requires about 540 calories per gramme of water. The primary source of energy is the sun.

The role of wind is two-fold. It carries water vapour away from the leaves, steepening the gradient (sink strength) or it may bring additional energy known as 'advective energy'. The advective energy can increase or decrease the consumptive use of water in a given locality. If there is a mass of active green vegetation, covering the ground completely and growing extensively under a non-limiting water-supply, the quantity of water evapo-transpiration by that mass of vegetation per day is independent of that species. Water loss by crop plants is, thus, primarily a physical phenomenon governed by the available solar and advective energy and the sink strength of the wind.

The consumptive needs of water by a crop cannot be reduced and no moisture economy is possible in the case of this item on a field scale, if the aim is to obtain the maximum yields. The entity B in the expression of water requirement comprises application losses during the conveyance of water through seepage and deep percolation. The irrigation efficiency depends upon the soil texture. The finer is the soil texture, the lower are the application losses and the greater is the irrigation efficiency than surface irrigation.

The management practices also influence the irrigation efficiency. The levelling of the land properly, the selection of the optimum plot size and the stream size and the proper scheduling of irrigation depth increase the efficiency. The application losses can also be minimised by using water-control structures, such as water gates, diversion boxes and by lining the canals. The entity C, i.e. special needs, depends upon the nature of the crop grown, the salinity status of the soil or water or of the both, the soil moisture status at the time of land preparation.

More water is required for crop production if the transplanting of the seedlings is involved, if the soil or irrigation water is saline or if the soil is dry and hard before preparing the land for sowing. The best, most rapid and most reliable method of finding out the consumptive use of water is with an open-pan evaporimeter which assesses all meteorological factors in an integrated way. The consumptive use of water is 0.7 to 8 times the value of the US class A open-pan evaporation when the soil moisture is adequate and the crop canopy is full and in an active stage of growth.

A sunken screened open-pan evaporimeter furnishes directly the values of consumptive use of water by crop plants during their active growth. The effeciency of field irrigation can be determined by measuring the quantity of irrigation water applied and stored in the root-zone. If it is not possible to measure the efficiency, the following broad values are suggested as a guide for surface-irrigation methods.

Soil class	*Irrigation efficiency (%)*
Sandy	60
Sandy loam	65
SandyLoam	70
Clay loam	75
Heavy clay	80

The quantity of water needed for special operations, such as land preparation, leaching down salts and transplanting, can be calculated by actual determinations in each case. The following values are indicated as a guideline.

Operation	*Details*	*Quantity of water (mm)*
Land preparation	Light soil	50 to 60
	Heavy soil	70 to 80
Puddling and transplanting of rice	Light soil	300
	Heavy soil	200
	Other crops	75 to 100
Leaching down of salts	Low salinity	10 per cent of the consumptive use
	Medium salinity	20 per cent of the consumptive use
	Heavy salinity 25 to 40 per cent of the consumptive use 75 to 100	

To find out irrigation requirements on a canal system, a period of 10 to 15 days is chosen. The values of irrigation needs during this period are worked out by subtracting the values of effective rainfall, ground-water and dew from the total water needs. These periodic values, when totalled for the entire period, furnish the total irrigation needs in a locality after making the correction for irrigation efficiency.

Irrigation requirement 'iw' during a period 'n' =
Total water needs during the period 'n' (A+C) – Contributions from effective rainfall, dew and ground water during the period 'n' x (Irrigation Efficiency %)

Total irrigation needs during the season = N epsilon l iw

In the case of rice, the percolation need are worked out with the drum-culture technique and added to the values of consumptive use to arrive at the value of the total water needs. Subtracting the values of effective rainfall, ground-water contribution and dew from the total needs, the net irrigation needs would be obtained, when divided by the irrigation efficiency, the values of gross irrigation needs will be obtained. In the case of the drum-culture technique, two drums, one with the bottom and the other without the bottom, are installed in the field.

These drums are filled with water maintaining its level equal to that in the outside field. Daily, the difference between the water levels in the drums with and without the bottom gives the values of seepage losses. Differences between the two daily consecutive readings in the drum with the bottom furnish the values of the consumptive use of water. A third drum with its bottom, but with an overflow device is also installed for knowing the effective rainfall. The differences between the water-level in this drum and that in the drum without the bottom furnishes the values of ineffective rainfall.

When to Irrigate

A crop should be irrigated before it receives a setback in its growth and development. There are several devices and methods, such as the use of blocks of Plaster of Paris, tensiometers, neutron moisture meter, soil-sampling, etc. for scheduling irrigation. But these devices and methods are either laborious and costly and are suitable only for experiments. A careful observation of the system on leaves is not of practical use in the case of field crops.

Plants growing on a slightly higher level or on a spot where the texture of the soil is relatively light in relation to the surrounding soil will show the symptoms of wilting earlier than the plants growing in the rest of the field and will furnish advance intimation for the application of irrigation. It is also possible to prepare a special spot of about one-to-two-metre cube by mixing a sand with the original soil. Plants growing on this soil-cum-sand mini-plot would indicate the irrigation need earlier than the rest of the field. This is the most inexpensive, simple and practical method suited for any crop and for any soil, except sandy soils.

How Much to Irrigate

Measurement of irrigation water: The application of precise quantities of irrigation water not only ensures a high efficiency of water use by the crops, but also reduces nutrient losses through leaching and results in better aeration of the soil. Irrigation water is measured under two conditions, viz. at rest and in motion. The units commonly used for expressing the volume of water at rest are litres, cubic metres, hectare metres, etc. The rate of flow is expressed in terms of units of volume per unit of time, e.g. litres per second, cubic metres per minute, hectare metres per day, etc.

The rate of flow of small streams on a farm can be directly measured by using a container or a bucket. The time needed to fill the container gives the rate of flow. This method can be used to determine the discharge rate of pumps and other water-lifting devices. The rate of flow of water into or out of a reservoir of known dimensions can also be directly made just by measuring the change in the water level of the reservoir with respect to time. Several types of water-meters are commercially available. They can be used for measuring the volume of water directly.

The rate of flow of water at a point in a straight ditch or in an open channel can also be determined by multiplying the cross-sectional area of water at that point by the average velocity. Under field conditions, it is impossible to maintain a uniform cross-section and, thus, a uniform velocity at various points. The other devices commonly used for measuring irrigation water on a farm are orifices, weirs and flumes of various predetermined shapes and sizes.

In most of these devices, the head of water on the upstream side is measured. The rate of flow is calibrated to it. Standard formulae, tables and calibration charts can be used to calculate the discharge of water through these devices. The commonest among these devices are the 90° V-notch weir and the Parshall flume.

Depth of irrigation: The quantity of water needed for net irrigation on different soil types per metre depth of soil profile at 50 per cent of soil-moisture availability is as follows :

Soil type	*Irrigation depth (mm)*
Sandy	25 to 50
Sandy loam	45 to 80
Sandy loam	70 to 110
Clay loam	80 to 120
Heavy clay	100 to 140

The depth of the soil to be wetted by irrigation depends upon the extent of the root-system of different crops. In the case of annual crops, the root-system is a dynamic one, growing to its full depth till about the flowering stage. In the case of perennials, the root-system is, more or less, stable in depth after about five years. The depth of irrigation, therefore, should be decided by taking into account the root-system of the crop. When rains are anticipated, it is better to under irrigate the field crop to accommodate the rainfall and reduce the amount of irrigation water to be applied. Important crops can be classified as under with respect to their maximum root-system on a normal well-drained soil.

Shallow-rooted	*Medium-rooted*	*Deep-rooted*	*Very deep-rooted (more than 90 cm)*
Nursery beds, Leafy vegetables, Onion, Garlic and Flowering annuals	Rice, Maize, Wheat, Millets	Lucerne, *Berseem*	Fruit-trees other than those mentioned here in the previous columns, forest species
	Groundnut, Lawns, Tobacco	Cotton, gram, Sorghum, Barley	
	Cereal forages, Vegetables, Papaya, and Banana	Sugarcane, Grain, legumes, Safflower, Castor, and Citrus	

Recent researches have indicated that a greater frequency of irrigation is more conductive to higher yields than a lower frequency. This means that it is advisable to apply light and frequent irrigations instead of heavy irrigations at longer intervals. The advantage from this practice is due to the increased uptake of nutrients, which are normally in the surface layers of thesoil profile, and also due to smaller losses through leaching.

The new strains do not need higher quantities of water during their entire life than the old ones. But the new strains do require timely irrigation, since their productivity per day is very high as may be seen from Table 4. In other words, the new strains are more sensitive to moisture stress and delayed irrigation and moisture deficiencies which hit them more severely, resulting in heavy losses in yield.

Table 5 shows what difference is made by four irrigations, timely and untimely, to a crop of wheat. The quantity of water applied (delta) in both the cases was nearly the same.

Similarly, fertiliser applications do not increase the potential water needs of a crop for all practical purposes. But they increase the water-use efficiency by enhancing production. Nutrients in fertilisers need to be dissolved in water. They must move down from the soil surface to which they are generally applied and be distributed in the root-zone.

Table 5. Productivity of old and new strains of crops

Crops	*Varieties*	*Approximate duration (days)*	*Grain yield (kg/ha)*	*Productivity (kg/ha/day)*
Rice	Old, tall	150	3,000	20
	New dwarf	150	4,500	45
Sorghum	Old	125	2,500	20
	New hybrid	110	4,500	40
Pearl-millet	Old	100	2,000	20
(*bajra*)	New hybrid	95	4,000	42
Maize	Old	110	2,200	20
	New hybrid/ composite	110	5,000	45
Wheat	Old, tall	140	2,500	18
	New, dwarf	130	5,200	40

Number of irrigations during crop age (days)			*Yield (q/ha)*	*Relative yield*
0-60	*60-90*	*90-120*		
—	2	2	37.0	100
1	1	2	44.0	119
2	1	1	51.3	140
2	2	—	48.5	131
2	—	2	52.3	141

Soil-moisture measurement: The knowledge of amount of moisture present in the soil is essential for deciding the time and the quantity of water to be applied. Several methods are known, and the choice depends on how precisely and quickly the information is needed, and on the availability of the requisite instruments.

Field methods: Field methods yield immediate information, but the accuracy is just satisfactory for practical purposes. A farmer can assess the moisture status in the root-zone by inserting an iron probe or a tube. Resistance for forcing down the iron probe depends on the moisture content of the soil profile. In a wet soil, the probe will be forced down more rapidly and easily than in a dry soil. If a hollow tube is used for soil-sampling, the moisture can be estimated from the colour and the dry soil and will easily form a ball. By experience, these methods can be used fairly precisely for scheduling irrigations.

Laboratory methods: Laboratory methods furnish precise values but involve costly devices and may be laborious and time-consuming. In the case of the standard gravimetric method, soil moisture is estimated by taking a sample with an augur and determining the loss in its weight when dried in an oven at 105°-110°C to a constant weight. The loss is expressed on the dry-weight basis. When there is no oven and electricity is not available, the soil sample can be dried by burning it with benzene, alcohol, or spirit and the loss in weight is recorded. This method, however, needs calibration with the gravimetric method.

The moisture percentage, expressed as percentage on the oven-dry basis, is converted into the total available moisture on volume basis by multiplying it by its bulk density. The bulk density is the dry weight of the soil per unit of its volume. Thus, if a soil has 20 per cent moisture on the oven-dry basis and weights 1.5 g per cubic centimetre, it contains 30 per cent water on volume basis or 30 cm per 100 cm depth of the soil when the area is the same. At research stations, a record of soil-moisture changes can be obtained instantaneously by using devices, such as a tensiometer or a resistance block, which are rather costly.

In research institutes, more efficient devices, such as the neutron moisture-meter or the gamma-ray method, can be used. These devices have a high precision but are very costly. A tensiometer measures the force with which the surrounding soil is pulling the water held by it. A simple tensiometer consists of a porous porcelain cup, connected with a tube to a mercury manometer or a vacuum gauge. The tube is filled with water and the tensiometer is installed in the field with its cup at the desired soil depth in the active root-zone of the crop.

When the soil moisture is depleted, water from the porous cup moves out, creating a vacuum in the system. The vacuum is recorded on the gauge or the manometer. With a multiple-cup tensiometer, the values in respect of the moisture status can be assessed with integration. A tensiometer works satisfactorily only in the wet range of available soil moisture.

A resistance block contains the electrodes embedded in a moisture-sensitive material, such as gypsum, nylon, and fibre glass. The block is installed at the desired depth in the soil and the wire leads are brought above the soil surface. When connected to a potentiometer, the block indicates the resistance to the flow of electricity, that varies with the moisture content of the matrix or the soil. In a wet soil, the resistance is low, whereas, in a dry soil, it is high. The device is calibrated against the gravimetric method for interpretation.

REFERENCES

Anderson., *Water Rights: Scarce Resource Allocation, Bureaucracy, and the Environment*, 1991.

Anita Roddick, *et al., Troubled Water: Saints, Sinners, Truth And Lies About The Global Water Crisis*, 2004.

Diane Raines Ward., *Water Wars: Drought, Flood, Folly and the Politics of Thirst*, 2002.

Marq de Villiers., *Water: The Fate of Our Most Precious Resource*, 2003.

Miriam R. Lowi., *Water and Power: The Politics of a Scarce Resource in the Jordan River Basin*, 1995.

William E. Marks., *The Holy Order of Water: Healing Earths Waters and Ourselves*, 2001.

5

Water Table Management

As water infiltrates through pore spaces in the soil, it first passes through the zone of aeration, where the soil is unsaturated. At increasing depths water fills in more spaces, until the zone of saturation is reached. This relatively horizontal plane atop this zone constitutes the 'water table'. A sustainable amount of water within a unit of sediment or rock, below the water table, in the phreatic zone is called an aquifer. The ability of the aquifer to store groundwater is dependent on the primary and secondary porosity and permeability.

FORM OF A WATER TABLE

The form of a water table may change and vary due to seasonal changes, topography and structural geology. In undeveloped regions, or areas with high amounts of precipitation, the water table roughly follows the contour of the overlying land surface, and rises and falls with increases or decreases in infiltration.

Springs and oases occur when the water table reaches the surface. Springs commonly form on hillsides, where the earth's slanting surface may "intersect" with the water table. Other, unseen springs are found under rivers and lakes, and account for the base-flow water levels in water bodies.

Surface Topography

Within an aquifer, the water table is rarely horizontal, but reflects the surface relief due to the capillary effect in soils, sediments and other porous media. When water reaches the zone of saturation the movement of the water is no longer vertical, it is horizontal in the direction of the slope of the water table.

The slope of the water table stand for the hydraulic gradient, which then depends on the rate at which water is added to the system and the permeability of the material. In hilly regions, the variation in gradient give rise to rivers, springs or oases when the water table intersects the surface. It should be noted that the water table does not always mimic the topography due to variations in the underlying geologic structure.

Perched Water Tables

A perched water table (or perched aquifer) is an aquifer that occurs above the regional water table, in the vadose zone. This occurs when there is an impermeable layer of rock or sediment (aquiclude) or relatively impermeable layer (aquitard) above the main aquifer but below the surface. If a perched aquifer's flow intersects the Earth's dry surface, at a valley wall for example, the water is discharged as a spring.

Fluctuations

Seasonal fluctuations

In some regions, winter precipitation is often higher than summer precipitation and so the groundwater storage is not recharged in summer. Consequently, the water table is lower in the summer period yearly. This disparity between the level of the winter and summer water table is known as the zone of intermittent saturation, wherein the water table will fluctuate in response to climatic conditions.

Long term fluctuations

Fossil water is groundwater that has remained in an aquifer for millennia, and occurs mainly in deserts. Fossil water is non-renewable by present day rainfall due to its depth below the surface, and any extraction (such as mining) causes a permanent change in the water table in such regions. There are some agronomic and water management practices which may be implemented at the field and watershed scales to reduce the concentration of chemicals in agricultural drainage water. Agronomic practices include tillage, cropping methods, and improved and timely fertiliser and pesticide applications.

One water management practice which has been shown to significantly reduce NO_3-N in drainage water is water table control. The technique has been tested in the humid regions of eastern Canada, and the eastern and mid-western United States. It is currently promoted in these regions as a combined drainage/irrigation practice for increasing crop yields and reducing NO_3-N pollution from relatively flat, artificially drained cropland. The quality of receiving waters is thus improved.

The water quality benefits of water table management are well described in Gilliam and Gilliam et al.. Water table management systems are currently being tested in some parts of Europe and Asia. The practice is yet to be evaluated in the arid and semi-arid regions, although there is potential for water table management in these regions. However, proper management of the water table is required to ensure that re-salinisation of the crop root zone does not occur.

Early attempts at sub-irrigation in arid and semi-arid regions failed because of soil salinisation. More information is required on the effects of different water table levels on crop yields and salt movement in the root zone in various climatic locations and soil types. In the humid regions of

North America, water table management has evolved on lands which were first improved with surface and subsurface drainage.

Drainage improvements were initially required on these poorly drained soils to increase crop productivity and provide better soil conditions for field machine trafficability. With concerns about the quality and impacts of agricultural drainage water on marine ecosystems, as well as nitrate contamination of drinking water, the drainage systems were modified to reduce the concentration of pollutants in the drainage water. Water table management has been practised for several years on organic and sandy soils in humid regions to reduce drought stress of high value crops.

Types of Water Table Management

There are two forms of water table management: controlled drainage and sub-irrigation. Controlled drainage restricts the discharge from a subsurface drain outlet or open main drain, resulting in a higher field water table. The water table drops naturally over time due to evaporation and deep seepage, and is only raised if there is precipitation. With sub-irrigation, water is pumped slowly and nearly continually into open ditches or a subsurface drainage system to maintain a near constant water table.

When large rainfalls occur and the field water table rises above the desired level, the irrigation pump is stopped. The excess water then drains from a control structure in the ditch or drain outlet. As a result of higher water tables, water is provided for plant uptake through capillary rise or upward flux. This water helps to meet plant transpiration requirements and reduce drought stress. Corn and soybean yield increases of 10-15% have resulted from sub-irrigation in eastern Canada. An added benefit of the higher water table is that it enhances denitrification, thereby reducing nitrate leaching.

Denitrification is a microbial process whereby nitrate, rather than oxygen, is used to reduce NO_3^- to N and NO_x. Nitrates retained in the soil profile are available for plant use. Several researchers, including Willardson et al., Bengtson et al., Gilliam and Skaggs, Evans et al. and Lalonde et al., have shown that water table management can reduce NO_3-N in drainage water by over 60%. Denitrification also occurs in water stored in ditches. Other benefits of water table management include: reduced downstream peak flows and drainage volumes, and conservation of soil water to supplement irrigation water requirements. There are, however, some instances where water table management may increase peak flows.

Water table control structures

In areas with open canal drainage systems, flashboard riser control structures may be used. These are installed in the ditches, and by either inserting or removing a flashboard, the water level in the ditch is either raised or lowered. These structures can also be used to store runoff in canals, which can later be used for sub-irrigation. Surface runoff from drained fields can also be controlled with the aid of the flashboard risers.

This adjustable water table control structure, attached to the collector outlet, has a vertical riser through which water can be pumped for sub-irrigation. With pumped drainage outlets, the pumping system can be designed and managed to maintain a desired water level in the sumps, and thus in the field. It is possible to automate the system using a water level sensor installed in a dipwell at the midpoint between subsurface drain laterals. The sensor can switch the pump on or off.

Benefits of Water Quality

Drainage Hydrology

Water table management systems may be designed to

control drainage volumes, peak flow rates or chemical losses from agricultural fields or catchments. Once the design objective has been determined, the potential hydrologic impacts and the resulting changes in water quality must be determined. Hydrologic impacts will depend on the degree and method of drainage improvement already attained in the proposed project area. High water tables will result in low infiltration capacities of the soils and high surface runoff rates, and will increase the loss of chemicals located at or near the soil surface. On the other hand, unrestricted drain discharge will produce low water tables and higher infiltration capacities.

Consequently, during periods of high drain discharge, there is increased potential for the leaching of soluble chemicals existing in the soil profile. The reduction of drainage volumes may be a first step towards improved drainage water quality. Evans et al., have indicated that water table management may reduce total drainage outflow by over 30% compared to conventional drainage systems without outlet control. Outflows, however, vary widely, depending on soil type, rainfall and type of drainage system. During very dry years, water table management may totally eliminate outflow. In wet years, it may have little or no effect on total outflow.

During the growing season, water table management typically reduces outflow by less than 15 %; much of the reduction occurs during the non-growing period, i.e., winter and early spring. The decrease in outflows is not due to a significant increase in ET. A slight increase may be observed during the growing season, but this would not explain differences during the winter, when most of the reduction occurs. It is possible that most of the water is lost by lateral seepage to remote sinks, or by deep seepage to groundwater.

Depending on the topography of the watershed, this water may reappear downstream and it should be quantified. Generally, with improved drainage, there is a tendency for the planting of higher value crops, and increased chemical applications. This may lead to increased chemical loads in surface runoff and subsurface drainage discharge. Crop production methods should, therefore, focus on more efficient water, fertiliser and pesticide applications. Water quality benefits associated with water table management will only be achieved if sound agricultural practices are followed.

Nutrient Concentrations

Nutrient losses in agricultural drainage water raise many environmental concerns. Eutrophication, due to high N and P concentrations, increases algal growth and reduces oxygen in surface waters. From agronomic and economic standpoints, nutrient losses represent a decrease in the efficiency of the crop production system. There are also human and animal health concerns, because nutrients, especially NO_3-N, can contaminate drinking water supplies. Concentrations exceeding 10 mg/litre can be harmful to humans.

Nitrogen, P and K are the three major elements of fertiliser and manure. Nitrogen is easily converted to NO_3-N. Table 1 shows the order of magnitude of total annual losses and peak concentrations of these three nutrients which may be expected in surface runoff and subsurface drainage. It can be seen that P and K are of less concern than NO_3-N. Nutrient concentrations in drainage water are influenced by many factors, including: rate of fertilisation; soil and crop type; and soil water regime during the growing season. These factors affect the efficiency of nutrient uptake by the crop roots, and residual nutrient levels in the soil during the fallow or dormant season.

Nitrogen

In most cases, water table management aims at reducing

NO_3-N leaching. When developing a water table management strategy, losses of N in its different forms should be taken into account. A decrease in one form of N may result in an increase in another form. This is particularly the case if a decrease in subsurface drainage discharge results in an increase of surface runoff. In such a case, there might be an important decrease in NO_3-N concentration in drainage water, while a small increase in loss of organic N and NH_4 may occur.

Table 1. Orders of magnitude of peak concentrations and annual losses of NO_3-N, P and K in drainage water at field level

	Losses	NO_3-N	P	K
Surface runoff	Peak concentration (mg/litre)	50	10	10
	Annual (kg/ha)	1-20	1-20	1-10
Subsurface drainage	Peak concentration (mg/litre)	200	10	10
	Annual (kg/ha)	1-100	0.1-1.0	1.0-10

Knowledge of the N cycle, and the impact of agronomic practices on this cycle, will help with the implementation of a water table management strategy. Temperature and the soil water regime influence nitrification and denitrification. At the field level, the potential risks must be assessed by comparing the water table regime to: (i) fertiliser rates and practices during the growing season; (ii) the periods of high mineralisation rate; and (iii) the periods of possible denitrification. These factors are illustrated in Table 2.

High water tables due to water table management may significantly decrease NO_3-N concentrations in drainage water. Annual losses of total N from fields with water table management could be 40-50% lower than for conventionally drained fields. This decrease is due to both a reduction in drainage discharge and an increase in denitrification. If controlled drainage is used during the early spring or late winter, when drainage is not needed for crop production, one can, if temperatures are warm enough, dramatically increase denitrification and reduce NO_3-N losses.

Table 2. Factors influencing the nitrogen cycle in agricultural soils

Factor	*Effect on crop*	*Effect on nitrogen cycle*
Temperature	Higher ET which enhances vegetative growth and thus nitrogen uptake.	Higher temperatures (in the range of 20-30°C) increase the rate of most biochemical processes, especially mineralisation and denitrification.
Soil water regime	High water tables restrict crop growth and nitrogen uptake by roots.	Soil water contents below field capacity increase the rates of all processes, except denitrification. Saturated conditions due to higher water tables enhance denitrification.
Fertiliser and agricultural practices	Application of fertilisers increases the rate of crop growth and development.	Drainage improvements are generally linked to changes in cropping systems, which may lead to increased N fertiliser inputs. The conversion of grassland to cropland results in mineralisation of N accumulated in the soil profile.

The design of a water table management strategy to control N losses should take the following into account:

i) The timing and amounts of drainage discharge, before, during and after the growing season.

ii) The potential denitrification. Denitrification requires low oxygen concentration, due to shallower water tables, as well as the presence and activity of bacteria. Temperature and soluble organic carbon content in the soil influence denitrification. Soil temperatures above 10°C are necessary to induce significant denitrification.

iii) N_2O is a by-product of denitrification. This is a greenhouse gas, which is of environmental concern, and may only be a problem on very large watersheds.

iv) The scale at which the water table management is implemented. At the watershed scale, there is the possibility of less vegetation in and along channels. This may yield higher NO_3-N levels at the field edge.

v) Continually high water levels in drainage canals and/or ditches can cause problems of ditch bank stability and also promote problems with beavers and other rodents.

Phosphorus losses

Phosphorus losses are closely tied to sediment loss. Surface runoff is, therefore, the primary factor influencing both sediment and P transport. Shallow water tables during high rainfalls could increase surface runoff, and hence P movement. However, at the watershed scale, controlled drainage has been shown to reduce P losses.

Pesticides Concentrations

Pesticides are generally transported with soil particles in surface runoff. According to Munster et al., pesticide concentrations are generally higher in surface runoff than in subsurface drainage water. However, Kalita et al., stated that high pesticide concentrations in subsurface drainage water could be expected if preferential flow occurs. The amount of pesticide in subsurface drainage water is typically less than 0.1% of the amount applied. The pesticide most studied is atrazine, or compounds of the same family. This is a herbicide for maize crops.

Results for aldicarb, alachlor and metolachlor are also available. The concentration of these chemicals is generally very low, in the range of 1-3 m g/litre. Although further investigation is needed, a water table management strategy that aims to reduce pesticide losses should preferably reduce the amount of surface runoff. The water table level should therefore be lowered if a rainfall occurs immediately after pesticide application. However, Michaelsen showed preferential transport of several herbicides being enhanced in drained plots in northern Germany. Most pesticides have a relatively short half-life, and therefore do not persist in the soil or water for long periods.

Operational Aspects

Water table management systems can be operated by farmers, water boards or water users associations. However, there is also a need for an institution to monitor water quality. Sometimes a compromise may be required between agricultural and water quality benefits. Depending on the type and severity of water quality constraints, water table management will seek an optimum for either agricultural production or environmental protection.

System constraints may be either physical, technical or human. The most critical physical aspects include topography, the existing drainage network, water availability and soil conditions. Detailed information on the design, installation and operation of water table management systems can be found in the ASAE Standards. Water table management requires greater farmer involvement than does conventional drainage.

The operation of subsurface drainage systems requires few management decisions. However, with water table management, operational indicators are difficult to see, as the response to outlet adjustments is not immediately evident in the field. Furthermore, the operational incentive may not be solely increased crop yield, but also improved water quality and increased water quantities for downstream water users. These factors must be clearly defined for system operators at the start of the project.

Catchment scale

There are two approaches to the implementation of water table management: the catchment level and the field scale. These two approaches are complementary. The scale of implementation depends on: water quantity and quality objectives; farming systems; regional hydrology; institutional arrangements; and topography. Water quality is linked to farming practices and is therefore to be controlled primarily at the field level. However, field systems may necessitate control structures at the catchment level where water quality control depends on a reduction of drainage volumes and sediment transport.

Location of the control structures within the catchment will depend on the discharge and drainage area above the project site. The upstream drainage area may be several thousand hectares. In this case, it may be more economical to control the water level in field ditches with smaller structures. This may cost less than larger structures on main channels. The existing drainage network is also an important consideration. The depth and width of the channels govern water storage capacity. If a dual controlled drainage/sub-irrigation system is to be implemented, channel storage capacity has to be sufficiently large to provide irrigation water.

Topography and soils

Water table management is best suited to relatively flat lands. Sloping lands would require many control structures, to provide uniform water tables. Land slopes of less than 1% are recommended. However, as pointed out by Evans and Skaggs, while there is no absolute limit, slopes of less than 0.1% are the most practical. At the watershed scale, a slope of 0.1% would require a control structure every 300-1 000 m of channel length. Soil permeability is a key factor governing the rate of water movement from ditches to adjacent fields, and the upward movement of water from the water table to the plant roots.

Good lateral movement of water will occur in moderate to highly permeable soils. While Evans and Skaggs recommended a permeability of about 0.45 m/day, experience in eastern Canada has shown that water table management is also feasible on soils with permeabilities of 0.3 m/day. A restrictive soil layer, not far from the bottom of the subsurface drain pipes, will reduce deep seepage losses.

Monitoring System

Generally, a monitoring system needs to be implemented, both for system management and to measure water quality changes. Monitoring involves several factors and has to be performed on a regular basis, both at the field level by the farmer and at the catchment level by the water authorities.

Water quality impacts

Actual and potential detrimental impacts of drainage water quality both within the watershed and downstream should be identified. These investigations will help determine the principal water quality constraints and the most sensitive periods for control. They are best conducted by water authorities. The data may be useful in guiding fertiliser application.

Water table levels

Water table management requires periodic monitoring of the water table at the midpoint between drain laterals or ditches. It is best done by the farmer. One observation well per field is the minimum recommended. For a given soil and drain spacing, the frequency of observation and adjustment of the control structure depend mainly on the weather and crop development stage. For example, when crops are in their early stages of development, a shallow water table may impede proper root development, and this could make the crop more susceptible to drought later in the year.

Research by Madramootoo et al. indicates that water table levels between 0.50 m and 0.75 m from the soil surface are appropriate for most crops. The response of the water table to rainfall and control structure adjustments may be slow, depending on the soil properties and drain spacing. It may take several years to fully understand the response patterns and successfully operate the system. Accurate weather forecasts and the ability of the farmer to make use of climatic data could improve system operation.

Assessing system efficiency

Water agencies should organise a medium or long-term monitoring programme of the impacts of water table management on the hydrology, water quality and crop performance in the project area. Some of the main parameters to be monitored are shown in Table 3.

Table 3: Monitoring requirements

Impacts	*Parameters to monitor*	*Monitoring interval*
Peak flows	Surface runoff, subsurface drain discharge, water levels in streams	1-hour intervals with automatic devices
Soil erosion	Sediment in surface runoff	Accumulation after a rainstorm, using sediment samplers
Nutrient losses	Nitrate, ammonium, phosphorus	Monthly, during the growing season
Pesticide losses	Pesticide concentrations	Monthly, during the growing season

REFERENCES

Debenedetti, P. G., and Stanley, H. E.; "Supercooled and Glassy Water," *Physics Today* 56 (6), p. 40-46, 2003.

Postel, Sandra., *Last Oasis: Facing Water Scarcity*, New York: Norton Press, 1997.

6

Groundwater Management

Groundwater management is generally difficult to implement. Effective strategies are constrained by lack of data, limited technical capacities and, above all, lack of political and wider social support for standard regulatory or related interventions. Rapid demographic, economic and environmental changes in many regions often render such interventions inappropriate. By the time sufficient data and information are available to support management and the appropriate institutional frameworks are in place, conditions have often evolved to the point where initially envisioned management strategies have little chance of success. Change is an ongoing process and in many situations society is adapting rapidly to constraints and opportunities as they emerge.

GROUNDWATER PROBLEMS

Responses to emerging groundwater problems need to be identified on a more strategic basis than in the past. While integrated approaches to groundwater management in aquifers that are crucial to key uses may be feasible, the viability of establishing effective management systems over the large aquifers at risk in many countries is significantly lower. However, in many cases, people are responding to emerging groundwater problems through a variety of coping strategies.

In addition, people are migrating or developing livelihood strategies that are not water intensive in some areas. These changes, the incentive for which is often unrelated to groundwater, could represent a major avenue for reducing the social consequences of groundwater overdraft and the pressure on the resource base. Adaptive strategies that build off existing social trends and responses to water problems may contain opportunities for effective responses that the standard water management lens misses.

Groundwater Management Capacity

The limited groundwater management capacity (including the political capital needed to implement regulations and similar interventions) needs to be concentrated on a limited number of strategic, high-priority areas where it can be effective. In other regions, strategies that help people to adapt may prove more successful. Simple early-warning indicators that help guide such strategic approaches may be more important than detailed groundwater studies.

Rather than attempting to identify global problems or apply a single broadly integrated management philosophy to all situations, there needs to be more focus on adapting groundwater response strategies to the array of factors that determine opportunities and constraints in specific situations. Water scarcity is recognised increasingly as a global concern and, within that broad concern, more attention is focusing on emerging patterns of groundwater overexploitation and their implications for the availability of water to meet human and environmental needs.

Food production and food security are among the more important of these needs. Reliable water supplies, particularly those from groundwater, are the lead input for increasing yields, reducing agricultural risk and stabilising farm incomes. As a result, strong arguments can be made that access to groundwater plays an instrumental role in food security. Water availability and reliability are linked closely to food security, but the equation linking water to food security is partial and the links are neither linear nor transparent.

The full equation is a function of the interaction between water access, production economics and the wider network of entitlements that water users and others have within society. It cannot be assumed that a one-to-one relationship exists between access to reliable water supplies for irrigated agriculture and food security. Under the definition, food security often depends more on the ability of populations to purchase rather than produce food. This is because global and national food distribution systems now frequently negate the impact of local production problems on the availability of food in the market.

As a result, the question of whether people have access to sufficient food when groundwater problems disrupt agricultural production depends heavily on whether they have access to a diverse array of alternative income sources or to reserve capital. It also depends on wider factors such as transportation systems and the ability of countries to purchase and distribute food available on global markets. All this implies that analysis of groundwater availability and reliability on a project or regional basis is by itself a poor indicator of the vulnerability of the global population to food insecurity.

Nonetheless, access to water, and particularly to highly reliable groundwater sources, does play an important role in food security in many cases. Access to reliable sources of water reduces the production risk. Farm incomes at both micro (farm) and aggregate (regional) levels are buffered from the effects of precipitation variability, drought or general water scarcity conditions. As a result, access to reliable groundwater supplies can ensure the income flow needed to purchase food as well as playing a central role in food production. Furthermore, particularly in remote locations within developing countries, irrigated agriculture constitutes the sole source of income that is available to rural populations.

As a result, there can be a direct link between water access and household or regional food security. However, this link is highly dependent on the specific situation. There is no inherent direct link between water and food security. While access to water is important in many situations, in other situations irrigated agriculture is only one of many income sources or available livelihood strategies. Consequently, while falling water levels, irrigation system deterioration, droughts and other direct indicators of water scarcity can serve as signals that food security may be threatened, the actual degree of threat will depend on a wide variety of context-specific factors. Water scarcity measures are warning signals, but they do not on their own indicate the emergence of food insecurity.

Water Availability and Production

The most direct and tangible link between groundwater conditions and food security is that of water availability to meet crop requirements. However, water availability in an aggregate sense has little meaning as crop production is heavily dependent on seasonal and interannual fluctuations in availability, including timing in relation to crop growth stages. Many crops are vulnerable to moisture stress at critical points in plant growth, and their yields can be reduced substantially even if adequate water supplies are available following periods of shortage. For example, water stress at the flowering stage of maize can reduce yields by 60 percent even where water is adequate throughout the rest of the crop season. Similar impacts on onions, tomatoes and rice have also been documented.

In addition to the direct impact of water availability on crop growth, assured supplies are a major factor in inducing investment in other production inputs such as labour, fertilisers, improved seeds, and pesticides. As a result, as the reliability of irrigation water supplies increases there is multiplier effect on yields. Taken with the inherent flexibility of groundwater abstraction, these characteristics of groundwater were a major contributor to the role of irrigation in the green revolution. Irrigated agriculture now contributes almost 40 percent of world food production from 17 percent of cultivated land.

Expansion of irrigation was the 'lead' input driving yield increases during the green revolution of the 1960s-70s and subsequent decades. As the most reliable source of irrigation water, a source that can generally be tapped when and in the amounts needed, groundwater played a particularly major role. As Repetto comments: *"The Green Revolution has often been called a wheat revolution; it might also be called a tubewell revolution."* To this extent, this turnaround hinged upon high-value crops and the ability to pay for energy costs.

Yields in groundwater-irrigated areas are higher (often double) compared to those in canal-irrigated areas. In India, the groundwater-irrigated area accounts for about 50 percent of the total irrigated area and up to 80 percent of the country's total agricultural production may, in one form or another, be dependent on groundwater. Similar patterns are also present in other countries. In China's Henan province, tubewells serve about 2 million ha, or 52 percent of irrigated lands. Parts of Mexico, including some of its most productive agricultural areas, are also heavily dependent on groundwater.

The role of groundwater is equally important in industrialised countries. For example, Barraque estimates that: *"irrigation uses 80 percent of all water in Spain and 20 percent of that water comes from underground... The 20 percent, however, produces more than 40 percent of the cumulated economic value of Spanish crops."* Recent findings from Andalusia, Spain, indicate that groundwater-irrigated agriculture is economically more than five times more productive and generates more than three times the employment in comparison with surface-irrigated agriculture. The role of groundwater is important not only through higher yields in normal water years. In an analysis of wheat cropping in the Negev Desert, Tsur estimated the 'stabilisation value' as being *"more than twice the benefit due to the increase in water supply"*.

In southern California, the United States of America, where surface water supplies are less variable than in the Negev Desert, the stabilisation value in agriculture is as much as 50 percent of the total value of groundwater in some cases. During the drought in California, the United States of America, in the early 1990s, economic impacts were minimal largely because farmers were able to switch from unreliable surface supplies to groundwater. The value associated with the flexibility of pumped groundwater supplies has been a further boost to agricultural productivity as it has allowed intensification and diversification of agricultural production in otherwise inflexible surface-irrigation schemes.

However, the presence of groundwater irrigation alone cannot be given full credit for the increased yields documented around the world. It needs to be seen as part of a complementary and mutually reinforcing set of inputs. Groundwater availability has enabled farmers to invest in complementary inputs that, in combination, have increased crop yields substantially.

As FAO notes: *"the response of crop to fertiliser is higher where supply of irrigation water is assured compared to rainfed conditions."* It is the reliability and flexibility of groundwater that allows farmers to take the risk of investing in fertiliser, but which also substantially increases their crop productivity. For example, fertiliser use in Pakistan is highest in areas supplied by both canals and tubewells and thus having a highly assured supply of irrigation water.

The total nutrient application in these areas is 68.85 kg/ha compared to 29.0 kg/ha in rainfed areas. These observations point to the dependency of crop yields on interactions within a dynamic agricultural system and the difficulty of isolating a single factor as the primary factor contributing to increased production. Nonetheless, the information available indicates the critical role groundwater has played in agricultural production over recent decades. The relationship between assured supplies of irrigation water, increasing yields and food production is now under stress.

According to Rosegrant and Ringler: "*the growth rate in irrigated area declined from 2.16 percent/year during 1967-82 to 1.46 percent in 1982-93. The decline was slower in developing countries, from 2.04 percent to 1.71 percent annually during the same periods.*" Yield increase rates are also declining, and projections indicate that this will continue in coming decades. Furthermore, in some local areas such as Sri Lanka and in the rice-wheat systems of India, Nepal, Pakistan and Bangladesh, yields have been stagnant for a number of years.

Although stresses on water resources are increasing and there is a logical link between water scarcity and yield stagnation, causal relationships between emerging water problems, yields and food production vulnerability are not proven. According to Ladha *et al.*, where yield stagnation is concerned: "*There is some evidence of declining partial or total factor productivity...The causes for the stagnation or decline are not well known, and may include changes in biochemical and physical composition of soil organic matter (SOM), a gradual decline in the supply of soil nutrients causing nutrient (macro and micro) imbalances due to inappropriate fertiliser applications, a scarcity of surface water and groundwater as well as poor water quality (salinity), and the buildup of pests, especially weeds such as Phalaris minor.*"

Furthermore, as Seckler and Amarasinghe note: "It is very difficult to project crop yields. ... The international dataset does not distinguish between yields on irrigated and rain-fed area: they are just lumped together in average yields." Water is only one factor affecting crop yields. Data available at the global level do not provide much insight into the relationship between yields on irrigated and rainfed lands, or enable conclusions about yields on areas irrigated by groundwater or on areas where groundwater depletion is occurring.

Recent evaluations of the implications of water scarcity for food security range from the optimistic to the pessimistic. For example, Brown contends that primarily because of impending water shortages in northern China, the country will have to import up to 370 million tonnes of grain per year to feed its population in 2025. This massive increase in imports could cause steep increases in cereal prices and disruption of the world market. On the other hand, analyses by FAO and the International Food Policy Research Institute (IFPRI) indicate that yield increases will be the dominant factor underlying growth in cereal production in the coming decades and that, in aggregate, production increases will be sufficient to meet demand.

The FAO report states that: "The overall lesson of the historical experience, which is probably also valid for the future, seems to be that the production system has so far had the capability of responding flexibly to meet increases in demand within reasonable limits." The core point in this discussion of the direct links between groundwater availability and food production is the role of interacting dynamic systems and the uncertainty inherent in predicting outcomes based on partial understanding of any one of them.

The role that irrigation, particularly groundwater irrigation, has played in increasing yields is relatively clear. Whether or not emerging water problems are a significant factor underlying the declining rate of yield increases or represent a significant threat to overall production levels is less well documented.

While the potential nature of such connections is clear in concept, available data and other evidence are insufficient to test the conceptual relationships. While it is essential not to dismiss the implications of groundwater overabstraction and water scarcity for food production simply because data are insufficient to prove them, it is equally essential not to ignore the wide range of other factors that could be playing equal or greater roles. Therefore, the first part of the equation linking groundwater and food production is clouded even before investigation of the larger question of the role groundwater plays in entitlements and food security.

Food Security

Food security is a function of three factors: availability; stability; and the ability of individuals to obtain access to food. As Sen and others have argued for famines in India, starvation is frequently due to the inability of individuals to purchase supplies that are readily available on the market and is not a function of food availability per se. The entitlement approach described by Sen "*views famines as economic disasters, not just as food crises.*" Sen indicates that the main interest in the entitlement approach probably lies in "*characterising the nature and causes of the entitlement failures where such failures occur.*"

Sen's approach may have particular relevance for analysing the impact of emerging groundwater problems on food security. Studies in the late 1980s highlighted the critical role that access to water, particularly groundwater, plays in poverty alleviation. Reliable water supplies are a foundation that enables farmers to afford access to a wide range of development benefits and can also enable farmers to diversify into other, often non-agricultural, income sources. These benefits are accessed through the improved yields enabled by the green revolution package of inputs. However, they carry a substantial risk because farmers must make investments in fertiliser, seed and other inputs in order to achieve them.

These investments, which are often made on credit, will be lost if water supplies fail. Consequently, any decline in access to groundwater could have a major impact on the economic condition of small rural farmers. As Burke argues: *"the expansion of irrigated agriculture in the 20th century has de-coupled the water user from the inherent risk of exploiting both surface and groundwater resources. The apparent reliability of storage and conveyance infrastructure and the relative cheapness and flexibility of groundwater exploitation offered by mechanical drilling have sheltered the end user from natural hydrological risk."* If substantial groundwater-level declines occur, short-term risk exposure may return to levels not encountered since the spread of irrigation. This risk is predominantly economic.

The economic dimension is also central to understanding the meaning of groundwater overextraction. Most discussions of groundwater overabstraction emphasise the distinction between economic depletion and the actual dewatering of an aquifer. Large-scale aquifers are depleted in an economic sense long before there is any real threat of physical depletion. The Gangetic basin may have 6 000 m of saturated sediment, but only the top 100 m or so are economically accessible for irrigation. Furthermore, wells owned by small farmers are generally shallow. In the context of poverty and famine, falling water tables will tend to exclude those farmers who cannot afford the cost of deepening wells long before they affect water availability for wealthy farmers and other affluent users.

Consequently, substantial declines in water levels are particularly likely to have a major economic impact on farmers with limited land and other resources. This impact will tend to be particularly pronounced during drought periods when large numbers of small farmers could simultaneously lose access to groundwater as their wells dry up. A more creeping problem would occur during non-drought periods as water-level declines undermined the economic position of small marginal farmers, forcing them onto already saturated unskilled agricultural and urban labour markets. The food security crisis in both these situations would be economic rather than related to foodgrain availability per se. Furthermore, whether there actually is a food security problem would depend as much on larger economic conditions as on groundwater availability and the economics of agriculture.

Role of Groundwater in Socio-economic Development

The specific role of groundwater in socio-economic development has been widely recognised for the past 50 years. The specific advantage of groundwater is its ubiquity, and this ubiquity presents particular problems in its management. The spatial and temporal distribution of groundwater occurrences has presented certain opportunities for the development of rural communities, urban centres and irrigated agriculture. It is the patterns of groundwater use that serve as a starting point for considering management options. For example, the patterns and management of groundwater and aquifer use in urban areas are distinguishable from those in rural areas. There are two distinct styles of use that exhibit and require distinct management solutions for each setting.

In many arid and semi-arid urban areas, local aquifers are often the water resource of last resort and also the ultimate pollution sink. In these areas, the range of services provided by underlying aquifer systems is usually more complex than that demanded by adjacent rural dwellers. The systems of rights in use differ markedly. Rural users generally abstract groundwater themselves through wells that they own and control. Urban users are one step removed with extraction being the responsibility of organisations such as municipal supply utilities. This difference has legal implications.

Rural users anticipate access in the form of direct abstraction from local aquifers. It is important that this access be protected in the form of a property right or a declared right in use. On the other hand, most urban dwellers and businesses anticipate municipal services derived from groundwater resources without any sense of real engagement with the resource. The rural pattern is characterised by a mix of dispersed potable, stock-watering and irrigation applications. The irrigation can often be locally intensive and crowd out other uses. Drawdown externalities are evident in many systems as shallow wells dry up and pressures are lost.

The urban pattern differs markedly as the aquifer systems are expected to provide reliable sources of potable water from dedicated well fields and also to absorb effluent. Groundwater rise in superficial aquifers underlying urban areas poses geotechnical and public health problems. Irrigated agriculture and municipalities are often perceived as the predominant users of groundwater; urban centres looking for low to moderate volumes of high-quality water, and irrigated agriculture looking for high volumes of raw water, being indifferent about quality unless the groundwater is saline.

In practice, urban centres are struggling to maintain the quality of local groundwater sources and urban development is expanding on recharge areas. Irrigated agriculture is using large volumes of high-quality groundwater in inefficient field distribution and drainage systems with attendant problems of salinity and waterlogging. The advent of mechanised boreholes has encouraged rapid depletion of deeper, non-renewable sources. While this pattern of groundwater use is unsustainable in terms of renewable water resources, planned depletion and the assumption of substitutionality are seen as valid development paths. What is much more difficult to resolve is the sustainable development of the apparently low-intensity use of shallow groundwater in rural communities. In fact, this use is highly intensive in terms of the available resource base.

In many semi-arid zones of the world, the locally exploitable groundwater resource is often depleted or drained by the end of the dry season. In addition, domestic use is not limited to potable water and personal hygiene use, but also to stock watering and small-scale irrigation. In aggregate, these uses are significant and key to the welfare and development of the bulk of the populations in developing countries. There is now increasing pressure for the specific role of groundwater in environmental systems to be realised and accounted for.

To this extent, environmental systems could be counted as a 'user'. The problem lies in finding a way to allocate between uses and the need to value the set of environmental benefits and compare them with extractive uses. This is not a trivial issue, and a range of standard cost-benefit analysis and multicriteria analyses have been deployed to try and reconcile these competing uses.

Groundwater Research and Development

The bulk of hydrogeological and groundwater development/protection research remains largely technocratic. The volume of social research related to groundwater use is small in comparison with the volume of research on groundwater flow and remediation. This trend is changing to the extent that the 2002 IAH Congress focused specifically on the human uses of groundwater. In addition, two cross-cutting programme components Friend and Help interact through their operational concept with

IHP VI and will support any joint UN-agency programmes. Other specific initiatives on groundwater quality protection and remediation will require the application of innovative and especially cost-effective methods to clean up pollution in soils and aquifers including:

— *in situ* bioremediation approaches;

— reactive barriers;

— source zone treatment;

— flow and transport modelling;

— site characterisation.

These activities respond to declared needs from the specialist groups to examine institutional and policy issues in tandem with the technical matters. However, there appears to be a reluctance to view groundwater resource management from the perspective of the de facto regulator, i.e. the individual user with a mechanised pump. This suggests a need to approach groundwater management as a socio-economic issue together with a sound technical perspective.

The institutions 'responsible' for managing and regulating groundwater resources need to focus on social mobilisation as a priority. This does not imply that the role of technical understanding of aquifer systems is reduced. Rather, it strengthens the case for enhanced understanding and communication of these systems within broader political and socio-economic frameworks. However, the fact remains that aquifer systems are difficult to characterise, monitor and regulate.

The work by Shah is one of the few attempts to analyse the political and socio-economic context of groundwater. However, the distributed nature of the resource necessitates managing a large number of diffuse, low-intensity investment decisions. If these investments in groundwater are to be realised, then it is contingent upon the investors to examine the sustainability of the resource and the regulation of its abstraction by the multiplicity of users. Therefore, project formulators need to be aware of the potential problems associated with groundwater development.

In many cases, these have less to do with volumes of abstracted water than with the changes in groundwater levels that the abstraction produces. Thus, resource assessments need to focus on establishing the local hydrodynamic limits of abstraction and waste disposal. Equally important is the understanding of the potential for managing or influencing the new patterns of use, patterns that are often highly dispersed and individualised.

Impacts and Responses

The impact of emerging groundwater problems can be intensely local or regional in effect. However, the types of management responses to address the particular circumstances may not have been adjusted to these patterns of use and impact. In particular:

- Large-scale, publicly funded tubewell development has tended to be supply driven.
- Legal and regulatory provisions at national level cannot be policed adequately.
- Forcing the relaxation of overexploited systems may be impossible even where the economic rationale is compelling.
- Regulation of the drilling industry may provide information but may also inhibit development in some circumstances.
- The enhancement of indirect recharge may work for shallow groundwater circulation, but recovery of deeper systems requires sophisticated injection and alternative sources of high-quality water.

Progressive attempts to encourage the formation of groundwater user associations are in their infancy in developing countries. The results of irrigation management transfer and water user associations are now being documented, but specific groundwater examples have yet to emerge in the literature. Large-scale initiatives such as those developed for the Ogallala aquifer in the High Plains in the United States of America may have achieved some impact in attenuating the rate of decline of regional water tables. However, they have come too late to keep the agricultural systems productive for more than two or three generations.

Social Perspective on Technical Information

The perception and understanding of hydrogeological processes among groundwater users appear to vary considerably. Equally, the levels of information required in order to prompt adequate management responses will vary with: the nature of the aquifer; the level of technology applied; and the nature of the media. In many cases, a simple message about overabstraction or pollution may be sufficient to prompt a collective response by the community of groundwater users, provided it is the right information. In the case of the Ogallala, the initial public perception about recharge of the aquifer was erroneous.

Only when it was too late did a more realistic perception plant itself in the minds of the users. Local knowledge about groundwater-level fluctuations in shallow systems is often sophisticated and involves customary regulation in times of dry-season, water-table recession. For example, such behaviour occurs with the linear alluvial aquifer in Yemen and the more expansive basalt aquifers in Eritrea.

Under these circumstances, although the social perspective may be realistic, it will not necessarily inhibit overabstraction. With shallow circulation systems, pumping local aquifers intensively before dry-season recession sets in may be a sustainable strategy in terms of local food security. However, this paradox begs several questions concerning:

— the level of precision required to understand and communicate hydrogeological processes;
— the degree to which patterns and intensity of resource use can be assessed;
— the degree to which risk can be assessed, given the levels of uncertainty and limitations of available data;
— whether current data lend insights into key questions of access and system dynamics.

It is not possible to answer all these questions. Much of the hydrogeological data that is currently collected has little real meaning in relation to the core issues facing users. The level data do not give much indication as to whether users have access to a reliable source of water supply. In addition, they are unlikely to provide much insight into the flow dynamics necessary to really 'manage' an aquifer.

Coping Strategies

The need to find practical management practices would suggest that it is important to examine: (i) people's coping (adaptive) strategies, i.e. what populations do when faced with groundwater scarcity problems (water harvesting, alternative livelihoods and demographic shifts); and (ii) the policy implications (drought relief, climate-change response, investment directions, institutional forms). However, to date, no comprehensive analysis with specific application to groundwater has been carried out. Some limited information with respect to the impacts of irrigation and poverty is being compiled for an FAO study but this will cover a mix of surface and groundwater irrigation systems.

The absence of an ability to 'manage' aquifers on a sustainable basis is no reason for inaction. Adaptive strategies that build on existing patterns of social change could represent an alternative to groundwater focused 'management' strategies. In order to identify potential opportunities for this type of strategy, it is necessary to gain a better understanding of:

— who is actually affected by emerging groundwater problems (equity, etc.);
— what individuals communities and local governments are actually doing to address such problems;
— the short- or long-term coping strategies individuals follow when faced with water scarcity;
— the patterns of social and economic change that affect groundwater use and management options but that are not in themselves a response to perceived water management needs, e.g. changes in community characteristics occurring in response to globalisation.

Better understanding of the above could help to develop a more realistic 'adaptive' strategy rather than a technocratic 'management' strategy for responding to groundwater problems. It should result in the identification of a wide variety of practical points for intervention.

BASIS FOR GROUNDWATER ACCESS AND ALLOCATION

Key elements to recognise in the valuation of groundwater are:

— The strategic value of groundwater (water that is located near 'high-value' uses such as urban or prime agricultural areas) as opposed to aquifers located in less strategic locations.
— Aquifers with high-quality water that is not vulnerable to pollution (an aquifer that can be reserved more readily for strategic uses may have a higher inherent 'value' than a more vulnerable one).
— The types of uses (aquifers that produce high-value environmental services, such as instream flows, may

have a higher inherent value than others where such services are absent).

In most situations, groundwater is a common property resource with extremely high use value. The term 'common property' refers to the status of groundwater as a resource to which all overlying landowners generally have access. The term is not intended to reflect the legal status of groundwater as a 'public', 'common' or 'private' resource.

In some societies, groundwater is or has been linked to land ownership. In others it is viewed as a 'common heritage' to which all should have equal access, at least for basic needs. These conflicting positions are enshrined in religious doctrines such as the Shari'a and in western legal concepts dating back to Ancient Rome. At the same time, rights of access to groundwater have generally been linked to land ownership. Thus, there is often an unclear distinction between the 'private' nature of groundwater rights and the 'public' ownership of the resource itself. This contradiction is brought to the fore by increased recognition of the need for water to be used more efficiently.

Market mechanisms can play a major role in achieving efficiency objectives, and more emphasis is now being given to the nature of water as an economic resource in global debates. This emphasis translates into initiatives to clarify water rights, encourage water markets and issue 'concessions' in some countries. However, the process may be stillborn if there is no recognition that water resources, by their very 'public' nature, require regulation and are not amenable to absolute free-market solutions. As these initiatives increase, tensions related to underlying ethical issues can also be expected to rise. If groundwater is viewed as a common heritage to which all have fundamental rights of access, the ethical basis of unregulated concessions or markets that allocate water depending on ability to pay becomes controversial.

Valuation Methods

Estimates of the value of groundwater have assumed great importance in the context of the massive investments required to avoid pollution, remediate polluted aquifers and control overdraft. Some objective basis, related to the total economic value of the resource (including wider social and environmental values), is required to determine how much to invest in specific situations. However, the available valuation methods are partial at best.

Techniques for quantifying the economic value of groundwater resources include:

— Contingent valuation, which essentially involves asking people how much they would pay to maintain the resource or services dependent on it under carefully specified conditions.

— Hedonic pricing, e.g. obtaining a measure of the value of groundwater through differences in the value of lands with and without access to it.

— Derived demand and production cost analysis, essentially estimating the contribution of water to profits within a given set of economic activities.

— Loss analysis, estimating the value of groundwater as equivalent to the total social costs incurred when drought or depletion constrain economic activity.

— Averting behaviour in which the value of groundwater is estimated by the investments made to avoid water shortage.

— Substitution, the value of groundwater as equivalent to the least cost alternative source of supply for meeting the same set of services.

Many of the above techniques have been discussed in detail for the United States of America and in the context of Jordan.

The above quantitative techniques do not capture adequately many of the qualitative values associated with groundwater, the full range of services and benefits that it provides. They also assume perfect knowledge of the groundwater systems in space and time and the precise impact of uses. Non-use and *in situ* values, such as the prevention of saline intrusion and the protection of the environment against low-frequency events, are particularly difficult to establish values for. For example, valuation of ecosystem losses due to changes in groundwater conditions is an inherently subjective exercise.

Some sections of society approach the environment from a spiritual and religious perspective and, in doing so, place essentially infinite value on the natural resource base that sustains environmental conditions. Other sections of society focus on specific economic activities when valuing natural resources. Weighing the relative importance to give to these different perspectives is in itself a subjective value judgement, and one that is often implicit in the valuation methodology selected. A similar ethical divide surrounds questions of ability versus willingness to pay.

Quantitative measures of willingness to pay established by any of the primary methods ignore questions of ability to pay. For example, the socio-economic status of respondents can colour their responses in contingent valuation surveys. Access to groundwater is often a critical factor in the quality of life for socio-economically marginal populations in developing countries. However, their ability to pay for maintaining groundwater resources is often minimal. How this contradiction can be resolved depends on whether groundwater is treated as a common heritage or as an economic good.

Beyond the ethical debate lie practical questions regarding the degree of social and technical understanding of groundwater on which value estimates are established. A basic problem concerns the interlinked and sequential nature of groundwater uses. 'Waste' from one use is often the primary source of water for a subsequent use. Tracing the chain of uses is a difficult task, as is valuing the contribution of groundwater to each one. Beyond this, it is often difficult to determine whether values reflect low-frequency but high-consequence events adequately, e.g. determining whether a value for groundwater in agriculture reflects the consequences of droughts that occur with frequencies of 20, 50 or 100 years. This relates closely to issues of information and understanding.

The general public may tend to view groundwater as an underground bowl replenished regularly by rain. Values derived from this viewpoint are unlikely to reflect the finite nature of many groundwater resources or their vulnerability to pollution. Similarly, it is difficult to determine the extent to which other forms of capital can substitute for groundwater should existing uses constrain availability. The limits of substitution may only become evident when a chain of sequential and interdependent uses is understood fully. In sum, the degree to which people understand groundwater and its contributions to economic, environmental and other services may well influence the value they place on the resource.

Reconciling Competition for Groundwater

The aquifer systems under the most pressure from competing uses are typically those where municipal users are vying for adjacent groundwater being used by agriculture. In many instances, these systems are in arid and semi-arid zones where surface water alternatives are not available, e.g. the North China Plain and the Quetta Basin in Baluchistan, Pakistan. Reallocation is occurring, but it is generally forced, either as deeper water-supply wells capture water away from adjacent irrigation wells or as irrigation wells are closed down.

The growing municipalities in the central section of the Huaihe Basin in China have been able to capture groundwater from peri-urban irrigation but not through any planned reallocation. Reallocation of groundwater rarely occurs through the consensual transfer of use rights and in these instances it could be argued that the application of valuation methods would achieve little. Irrespective of the technical ability to capture groundwater, irrigated agriculture cannot compete with higher value municipal uses on economic grounds alone.

Concern has been voiced about the decline in water tables in key grain producing areas of the world. Brown cites China, India and the United States of America. However, others are less pessimistic. Custodio and Price call for more precise information on the hydrodynamic state of the aquifers in question. However, this is not to say that the depletion of key aquifers in major grain producing areas is not significant.

In areas of northern India and China, the resulting adjustment of local agricultural systems is causing migration from the land and social disruption. Equally, the long-term impact of aquifer drawdown is affecting surface flows and seepage, resulting in the loss of locally important wetlands. However, most of this evidence is anecdotal. Efforts to quantify the dependency of food security on groundwater using available national data can only be partial. Other more detailed attempts have illustrated the local problems with matching groundwater data and food production.

The concern that the reliance on groundwater irrigation is threatening global food security may be overplayed. Land irrigated by groundwater goes in and out of production incrementally as agricultural systems and markets respond to natural resource limits. The possibility that a shortfall in China's grain requirements would suddenly soak up the international market in traded grain may be remote, but what is troubling is the compounded effects of long-term drawdown and drought. In periods of recurrent drought when anticipated rainfall inputs are substituted by groundwater, the 'spike' in groundwater demand can result in sharp reductions at regional level. For example, stream flows in the western United States of America have been reconstructed based on dendrochronology.

Over the last 300 years, the average drought has lasted 5 years and some have lasted 15 years. Water levels in the Ogallala aquifer have been declining during periods of 'normal' or above average precipitation and the western United States is facing one of its most keenly felt periods of drought. Therefore, global food production problems could conceivably occur if there were simultaneous long-term drought across several grain producing regions where groundwater levels have already been declining. Irrespective of this, groundwater depletion, particularly in shallow aquifers, is already having direct socio-economic and environmentai effects. Some of these impacts may be lagged over several years or decades as aquifer hydraulics and hydrochemistry adjust to intensive pumping regimes. However, global analysis reveals little of the inherent tensions and opportunities that are experienced with groundwater irrigation on local scales.

LIMITS TO GROUNDWATER MANAGEMENT

An analysis of current practices might lead to the conclusion that there is no effective system of groundwater management. It is a rare exception when wells are closed down and capped off to prevent abstraction, or limits set on pumping durations or volumes. It is not such a rare exception to observe local, consensual enforcement of pumping limits, e.g. for irrigated agriculture in Eritrea and Yemen when pumped groundwater is rationed during the dry season. These arrangements are 'customary', but have only been occasioned by the advent of inexpensive motorised pumps.

The variable patterns of groundwater use and the varied services that aquifer systems provide do not form a clear aggregate picture or status of groundwater, nor do they present an opportunity for systematic management response. Despite the highly technical work that is carried out and presented in the hydrogeological literature, the status of knowledge of the aquifer systems is often limited at the level at which a management response is required. Highly detailed studies in contaminant transport are carried out in high-value settings. However, accurate and reliable monitoring and regulation in the crucial aquifers of northern India and Baluchistan, Pakistan, are not available. Even if they were, it is questionable as to whether the data would provide an effective tool for regulation or furnish a clear message for the education of users and the basis for behavioural change.

Implications for Integrated Management

One major concern is the fact that the issues outlined above are a symptom of current water management as a whole. In general, these management practices continue to ignore the integrity of groundwater systems even in arid regions where groundwater is the 'lender of last resort'. This is particularly true in the case of large sedimentary aquifer systems that are decoupled from contemporary recharge and are effectively non-renewable. In addition, the varying scales at which groundwater systems occur and are developed or exploited pose particular management challenges beyond those of conventional surface water or river basin management.

Therefore, a clear articulation of the specific guiding principles in groundwater development and criteria for evaluating policy responses to groundwater depletion and degradation is warranted. Such principles may have little to do with the more conventional principles of 'integrated water resource management', which are generally predicated on hydraulic control and regulation over river basins. This 'engineering hydrology' focus on water management continues to colour water resource management styles, which remain largely centralised and technocratic. One example is Namibia, where groundwater provides some 60 percent of bulk water, yet the institutional arrangements and investments have focused on the development of intermittent surface flows.

River basins and surface water irrigation schemes present 'neat' arrangements. The resource is naturally integrated at any point in the basin's watercourse and it is easy to monitor measurements, diversions, storage and abstractions. The impacts of irrigation abstraction and return flows occur in near real time and are immediately apparent upstream and downstream through hydraulic continuity. The system is neatly bounded, there are clear solutions of continuity, and systems of rights in use are generally clearly established. The same cannot be said for aquifer systems and groundwater development. Aquifer systems are known imperfectly, there are no clear solutions of continuity, responses are highly non-linear and can be lagged over centuries with none of the clear 'water year' rhythm observed in surface basins.

The largely technocratic, vertically integrated basin management models built around surface water schemes and the sets of incentives to surface water managers and users differ markedly from the management 'models' and incentives associated with the more imprecise character of aquifer systems and groundwater use. In addition, while the 'client base' for a basin manager would typically consist of a set number of well-identified user groups (irrigation schemes, water user associations, municipalities, etc.), the manager of an aquifer system may in practice have to engage with millions of individual users.

Therefore, the transaction costs of applying 'cross-sectoral integrated water resources management' in a classic sense can be expected to be several orders of magnitude higher. This does not appear to augur well for progressive conjunctive use management as a means to reconcile competing surface and groundwater demands. While conjunctive use is common in many irrigation schemes where individual irrigators have wells in the surface command area, the specific management of the combined sources is less common and mechanisms for doing this appear to be lacking.

However, progress in applying conjunctive use and the use of extensive informal water markets in and around surface-water and groundwater irrigation schemes demonstrate the ability of irrigation end users to adapt. Such adaptation often occurs in the face of contradictory signals and incentive structures established by higher order 'managers'. Whether such de facto arrangements offer more or less equity and economic 'efficiency' is debatable. Where regulation is weak or absent, the opportunity for the richer members of a groundwater user group to capture through technology or access to land will always be there. However, it is also possible to observe enhanced equity and efficiency through the myriad of small water, energy and pumping transactions that occur among irrigation user groups.

The sometimes unclear linkages between groundwater and food production have constrained the scope for management of the resource. Given the levels of uncertainty associated with groundwater information, the broader questions that remain are whether: (i) groundwater is amenable to the same types of management approaches associated with surface water irrigation; and (ii) it sits well within the frame of so-called 'integrated water resource management'. It is important to resolve these questions because the aquifers that are being used intensively can be expected to be in arid and semi-arid zones. Moreover, in such areas, surface water alternatives are scarce or unavailable and drawdown and pollution externalities can be expected to affect users within and outside the zone of groundwater use. Therefore, solutions that require some sort of 'integrated' or consensual and expert management will become imperative. However, requests to individual users to sacrifice private opportunity for the sake of basin or aquifer-wide efficiencies and equity are likely to encounter resistance, particularly where groundwater access is the principal means for accumulating assets and escaping from poverty.

In irrigated agriculture, the scope for addressing such management problems is also conditioned by the need to serve several policy 'masters'. Irrigated agriculture is a key component of many national agriculture strategies and a factor in the political equations many national leaders face. However, it is also expected to conform to water and environmental policy initiatives. That management of groundwater use for irrigation should be part of a national and regional commitment to integrated water resource management is not in dispute, but precisely how and through which policy instruments it should be effected often remains unclear.

In terms of irrigation practice, efficiency gains are important at all scales if pressures are to be reduced on environmental flows and downstream/downgradient users. Irrigation is not in a position to foreclose on other users. It would be simplistic to assume that irrigation efficiency is not significant at basin level. Irrespective of the impact across a particular basin or catchment system, it is the immediate deprivation of opportunity that may count. Similarly, it should also be appreciated that seeking efficiencies involves specific groups of water users and managers at the various levels. However, there may be no incentive for direct users to make efficiency gains if upstream managers cannot ensure conveyance efficiency.

With groundwater, this may not apply as the incentive is generally internalised entirely by the user.

Policy Response and Institutional Adaptation

In some cases, it is already too late to talk about the sustainable development of groundwater because the aquifers are already depleted, polluted or salinised beyond the regenerative capacity of their natural hydrogeological regimes. Some industrialised countries are moving toward a re-examination of groundwater management in a broader political and social context. Others are maintaining a more technical perspective. Elsewhere, many developing countries that rely on pumped groundwater to sustain agricultural output and supply municipalities continue to permit the intensive levels of abstraction with little evidence of pro-active groundwater management being deployed.

Technical regulation, economic incentives and participatory management approaches may offer the means to address groundwater management in the common interest. However, the character of initiatives will be determined necessarily by the local realities of the groundwater occurrences and the associated groundwater economy. Dealing with such diversity involves a different order of adaptability and flexibility than that normally associated with surface water or river basin management.

By the time groundwater arrives at the well head and enters irrigation ditches or raw-water pumping mains, it is perceived that groundwater management per se ceases and conventional water management takes over. Indeed, groundwater tends to be treated as the ultimate source of relatively high-quality water and the ultimate sink of used water. This occurs without any real appreciation of groundwater's regenerative capacity and its buffering role in the hydrological cascade. The management of surface water has fundamental implications for both groundwater quantity and quality at all stages and points of the hydrological cycle.

Therefore, it is essential to examine the scope for groundwater management not only in the strictest sense, but also as a prerequisite for integrated water resources management. This involves appealing to individual groundwater users in ways that have to do with advocacy and demonstration. It is significant that some political scientists sense a continuing tension between the 'eminence' of the State and the customary user rights of the beneficiaries. How this tension is resolved is critical for groundwater because the management of diffuse abstraction is highly dependent upon the approach local communities take to negotiating the use of a common property.

Historically, it has appeared easier, though not necessarily beneficial or cost-effective, for the State to control surface water abstractions and disposal within a vertically 'integrated' river basin plan where a central authority undertakes all the operational and regulatory functions related to the water cycle. Here the purpose of the State's intervention is the protection of the broader public interest in the basin's resources. However, this tendency in river basin management risks ignoring the important but highly distributed physico-chemical and socio-economic buffering roles of groundwater.

The approach may also rely heavily on regulatory measures as opposed to economic incentives to achieve desired results. More significantly, the array of users with which basin agencies tend to engage may differ markedly from the groundwater 'stakeholders'. Millions of individual farmers are not necessarily amenable to the same degree of association that can be recognised among urban utilities, industries and command area authorities. Thus, it can be argued that integrating groundwater use within both a physical and a socio-economic framework becomes not only an environmental necessity but also a political imperative where policies of decentralisation and subsidiarity are adopted.

Under these circumstances, the choice of water resource management instruments to address specific groundwater issues may still fall within the bounds defined by technical regulation, economic incentives and participation. However, they can be expected to take on a markedly different character from those associated with conventional surface water management. Equally, the quality of the information about aquifers, groundwater and user behaviour will also need to assume a specific character.

Gaps in Management

Groundwater acts as the primary buffer against the impact of climate variability and spatial variability in drought. However, as human development has become more susceptible to such variability, three major gaps in groundwater management have emerged, each with significant implications for sustainable development:

— The inability to cope with the acceleration of degradation of groundwater systems by overabstraction, and effective resource depletion through quality changes (pollution, salinity).

— In general, a lack of professional and public awareness about the sustainable use of groundwater resources. In particular, a lack of coherent planning frameworks to guide all scales of groundwater development and the consequent lack of appropriate policy responses and institutional development to prevent and attenuate degradation to groundwater systems.

— The failure to resolve competition for groundwater and aquifer services between sectoral uses and environmental externalities.

These specific concerns hinge upon the central issue of awareness. This relates as much to the groundwater related environmental concerns in industrialised countries as it does to the peri-urban communities in developing countries who depend on locally available groundwater sources.

In this sense, groundwater management regimes may be expected to encompass a set of economic, regulatory and ethical levers that are operated by markets, regulators/state institutions and user associations. Effective institutional approaches need to be aware of these socio-economic realities surrounding groundwater use. They also need to appreciate the inherent risks associated with development, the level of uncertainty (plus limitations in data quality) and the range of social pressures.

Filling the Gaps

In contrast to technical views of groundwater management, an overview of the Indian experience makes the point that although indirect demand management strategies, such as rice export controls in the Punjab, do not form part of the water management debate, they may ultimately have more impact on slowing the rates of depletion than equity-based institutional solutions. This begs the question as to whether there are practical approaches for responding to groundwater problems and their socio-economic impacts that are absent in current management styles. The strategies outlined below are starting points for expanding conventional management thinking by adapting to real points of socio-economic leverage and to aquifer and user-group scales.

In general, it is possible to observe the following characteristics of current groundwater 'governance':

— Lack of data and scientific understanding limit the ability of society to predict aquifer functioning and to develop realistic rights systems.
— Rights systems are difficult to design and implement

in most situations for a variety of technical and economic reasons.

- In most cases, social acceptance of private rights may be problematic.
- Aquifer management is politically complex because it would require active modification of established use patterns.
- The dynamic nature of both socio-economic globalisation and global climate change makes management complex. People are increasingly mobile and often have little incentive to participate in long-term management initiatives.

From among this set of characteristics, two broad types of management approaches for groundwater emerge: (i) 'thin and wide'; and (ii) 'thick and deep'. 'Thin and wide' approaches may encompass blunt tools such as power pricing, subsidies for efficient technologies, economic policies that discourage water intensive crops, etc. They can be applied over whole countries or regions. 'Thick and deep' approaches deal with specific aquifers on the basis of command and control management whereby aquifer management targets are set and enforced through a resource regulator.

Under these circumstances, groundwater management may be most realistic when applied in a limited manner to strategic aquifers for which a social consensus supporting management exists. This can be said to have happened in the case of the Qa'Disi aquifer in southern Jordan and the sister aquifers in Saudi Arabia. However, such examples are rare. Concerning recharge initiatives in India, here a 'thin and wide' approach to resource management may ultimately prove more successful than a 'thick and deep' approach to aquifer management, such as the technocratic initiatives to aquifer recharge along the Batiah coast in Oman and in the Quetta Basin in Baluchistan, Pakistan. In any event, the impacts of groundwater management approaches require specific monitoring and evaluation periods in order to make an accurate assessment of success in terms of aquifer response alone. It is doubtful whether enough time has elapsed to allow such an assessment of the various initiatives and to identify clear bright spots. These considerations apart, focusing management on strategic aquifers would also allow society to concentrate the required scientific, monitoring and enforcement tools on relatively small areas.

In addition, any current users displaced by management could be absorbed more easily than if management were attempted over larger areas. The design of implementation strategies to initiate and then propagate across the area of concern will be key. The recharge movement observed by Shah will deserve attention where it is clear that such viability may have more to do with social structures than the technical feasibility of conservation and regulation.

In order to start addressing gaps in management, it is important to recognise that institutional innovation and adaptation will need to be more sensitive to the range of influences and management instruments. A diagnostic to develop such adaptations will need to cover:

— macroeconomic policies;
— sector policies;
— rights systems;
— institutions and capacities;
— regulatory frameworks;
— public involvement.

Against the 'soft' institutional strategies, it is possible to define sets of technical options that relate directly to groundwater. Although these options may present expanded opportunities to manage groundwater, they would have to be applied strategically in circumstances that are amenable (where uptake of technical strategies will succeed). Such technical options include:

— conjunctive management (conjunctive use and aquifer storage and recovery);
— conservation enhancement and protection;
— water harvesting and supply enhancement;
— irrigation efficiency improvement and demand management.

The design of a suite of institutional and technical strategies and their implementation at the scale required to make an impact (to conserve or reallocate groundwater resources) is unlikely to be achieved in the short term. Countries reliant upon groundwater irrigation, such as Mexico, are now considering periods of decades. It is up to agriculture to adapt and improve its water productivity and so release the pressure before the resource base is exhausted. In part the agriculture sector can achieve this through improvements in crop yields, irrigation efficiency and post-harvest processing. However, it may be non-technical interventions that play a greater role.

Resource substitution through food imports and transfers to higher value uses are already making their mark. Setting up the socio-economic tools for more flexible allocation of scarce resources will be pivotal to addressing equity and efficiency concerns. For example, improved systems of land tenure and water use rights can help considerably in spreading local water risk and improving food security by optimising local food production patterns.

Equally, the economic signals to food producers need to be clear and stable. Indeed, sound agriculture policy can set a strategic balance between rainfed and irrigated agriculture where there is effective demand and real comparative advantage in domestic resource costs. Furthermore, agriculture policies that encourage different forms of agriculture in different areas could be critical. For example, India has nationwide policies governing the support price for rice and wheat. Such national-level policies could be tailored to reflect regional water availability differences. Thus, they could encourage production of water intensive and less water intensive crops in areas with different comparative advantages. Here again, 'thin and wide' approaches can complement 'thick and deep' approaches.

This is where there is real scope for savings in agricultural water demand. However, it will involve a change in the balance between rainfed and irrigated agriculture and a number of related socio-economic changes. People are making social transitions as a result of many other factors that are not related to water resource availability or management.

Where people are making such a transition and the groundwater resource base is not being used in a sustainable manner, policies can be tailored in ways that encourage transitions towards low water-use livelihoods. This may not be part of conventional approaches to groundwater management, but it could be key where direct enforcement of water management targets is deemed impossible in either technical or political terms.

REFERENCES

Franks, F (Ed), *Water, A comprehensive treatise*, Plenum Press, New York, 1972-1982.

Gleick, Peter H., *The World's Water: The Biennial Report on Freshwater Resources*, Washington: Island Press, November 10, 2006.

Marks, William E., *The Holy Order of Water: Healing Earth's Waters and Ourselves*, Bell Pond Books, Great Barrington, MA, November 2001.

7

Management of Freshwater Resources

Preserving the quality and availability of freshwater resources however, is becoming the most pressing of many environmental challenges on the national horizon. Perhaps, because water is considered a cheap readily available resource, people fail to realise just how much stress human demands for water are placing on natural ecosystems. The stress on water resources is from multiple sources and the impact can take diverse forms.

The growth of urban megalopolises, increased industrial activity and dependence of the agricultural sector on chemicals and fertilisers has led to the overcharging of the carrying capacity of our water bodies to assimilate and decompose wastes. A deterioration in water quality and contamination of lakes, rivers and ground water aquifers has therefore resulted.

The need is to bring about a perceivable shift in our philosophy and address water problems. Water resource managers and professionals are beginning to focus on the use of existing infrastructure to meet the demands of a growing population by improving efficiency, reallocating of water for different uses, prioritising the water demand sector-wise and adopting policies and practices that check resource degradation.

PRESSURE

Uneven resource distribution

India receives an average annual rainfall equivalent of about 4,000 billion cubic metres (BCM). This only source of water is unevenly distributed both spatially as well as temporally. Most of the rainfall is confined to the monsoon season, from June to September, and levels of precipitation vary from 100 mm a year in western Rajasthan to over 9,000 mm a year in the northeastern state of Meghalaya.

With 3,000 BCM of rainfall concentrated over the four monsoon months and the other 1,000 BCM spread over the remaining eight months, our rivers carry 90% of the water during the period from June-November. Thus, only 10% of the river flow is available during the other six months. National level statistics for water availability mask huge disparities from basin-to-basin and region to region. Spatially, the utilisable resource availability in the country varies from 18,417 cubic meters in the Brahmaputra valley to as low as 180 cubic metres in the Sabarmati basin.

Even within the Ganga basin, water availability varies from 740 cubic meters in Yamuna to 3,379 cubic meters in the Gandak. Rajasthan, for instance, with 8% of the country's population has only 1% of the country's water resources while Bihar with 10% of population has just 5% of the water resources. Thus, while India is considered rich in terms of annual rainfall and total water resources, its uneven geographical distribution causes severe regional and temporal shortages.

Declining resource availability

Surface water

Of 4,000 BCM of available water from precipitation, the mean flow in the country's rivers is about 1,900 BCM. Out of this, only 690 BCM is utilisable. With 177 BCM of live storage created by the existing major and medium projects and another 75 BCM of storage from projects under construction, there is still a gap of 440 BCM of water, which is utilised.

Groundwater

The latest assessment of replenishable ground water resources has been made at 431.9 BCM by the Central Ground Water Board through a large volume of hydrologic and related data. This is the sum total of potential due to natural recharge from rainfall and due to recharge contributions from canal irrigation. The utilisable ground water resources have been assessed at 395.6 BCM. The CGWB has also assessed the quantum of static ground water resources at 10,812 BCM. Water availability from other sources and through desalinisation of sea and ground waters is considered negligible in view of the high cost.

The assessed gross available and utilisable water resources of the country, based on conventional technology, are therefore 2,384 BCM (billion cubic metres) and 1,086 BCM, respectively. With an estimated population of one billion in 2000, the available and utilisable water resources per capita per year are 2,384 m^3 and 1,086 m^3 respectively against an estimated availability of 6008 m^3 in 1947. This itself, gives a broad indication of the growing resource scarcity in India in the fifty three years since independence.

Resource Demand

Since independence, India has witnessed an unprecedented increase in population. From a population of about 343 million in 1947, the population has grown at a rate of 2.04% to cross the 1,000 million mark in 2000. With an increasing number of mouths to feed, there has been an additional pressure on agriculture resulting in an increase in net sown area from 119 million hectares in 1951 to 142 million hectares in 1997; high cropping intensity has also resulted in an increased demand for water resources.

Domestic water need in the urban areas has also grown notably with the current urban population at 4.5 times the population level in 1950s. The water requirement of the manufacturing sector has increased in proportion to the increase in the sector's share in GDP from about 12% in 1950s to 20% in 1990s. Further, there is a substantial variance in the different user sectors—agriculture, domestic and industry, vis-à-vis their share of water demand, resource pricing structure and usage efficiencies, which creates inter-sectoral competitions and conflicts.

The agriculture sector, for instance, accounts for about 95% of the total water demand with the subsidised and free regime of supply of power and water resulting in the over-exploitation and inefficient usage of water. The high resource cost for industries, on the other hand, cross-subsidises the water consumed by the other sectors. The demand for fresh water has been identified, as the quantity of water required to be supplied for specific use and includes consumptive as well as necessary non-consumptive water requirements for the user sector. The total water withdrawal/utilisation for all uses in 1990 was about 518 BCM or 609 m^3/capita/year.

Estimates for total national level water requirements, through an iterative and building block approach, have been made for the years 2010, 2025 and 2050 based on a 4.5% growth in expenditure and median variant population projections of the United Nations. The country's total water requirement by the year 2050 will become 1,422 BCM, which will be much in excess of the total utilisable average water resources of 1,086 BCM. At the national level, it would be a very difficult task to increase the availability of water for use from the 1990 level of approximately 520 BCM to the desired level of 1,422 BCM by the year 2050 as most of the undeveloped utilisable water resources are concentrated in a few river basins such as the Brahmaputra, Ganga, Godavari, and Mahanadi.

Inequitable water supply

Although about 88% of the population, on average, has been covered with organised drinking water supply in class I cities and class II towns of the country, there is a huge disparity in quantity of water supplied. Of the 299 class I cities, only 77 cities have cent percent water supply coverage. The per capita water supply also ranges from as low as 9 litres per capita per day (lpcd) in Tuticorin to as high as 584 lpcd in Triuvannamalai. Similarly, 203 of the 345 class II towns have low per capita supplies of less than 100 lpcd. Besides an inequitable distribution of water in a given city, the supplies are erratic with water quality degrading continuously over time.

Efforts have been made to provide safe drinking water in rural areas by providing funds in the state budgets from the First Five Year Plan onwards. The Central Government assists the states through programmes such as the Accelerated Rural Water Supply Programme (ARWSP) and Rajiv Gandhi National Drinking Water Mission (RGNDWM). The status of coverage of habitations as on 1 April 1999, shows that about 267 347 habitations (18.6% of total) are still either not covered or partially covered. Cent percent rural habitation coverage was only in Uttar Pradesh, Delhi, Pondicherry and Chandigarh. Percentage habitations fully covered in Assam, Punjab and Kerala are only 57.4%, 33.3% and 22.2% respectively.

Resource degradation

Discharge of untreated domestic wastewater is a predominant source of pollution of aquatic a resources in India. Urban centres contribute more than 25% of the sewage generated in the country. The smaller towns and rural areas do not contribute significant amounts of sewage due to the low per capita water supply. Wastewater generated normally percolates into the soil or evaporates.

The CPCB conducted a survey in 1994- 95 on water supply and wastewater generation, collection, treatment and disposal in 299 class I cities and 345 class II towns of the country. The survey findings indicated that, most cities did not have organised wastewater collection and treatment facilities. It also emerged that the class I cities and class II towns of Maharashtra, Delhi, Uttar Pradesh, West Bengal, and Gujarat contribute 59% of the total wastewater generated in the country.

Furthermore, the facilities constructed to treat wastewater do not function properly and remain closed most of the time due to improper design and poor maintenance, together with a non-technical and unskilled approach. Rural sanitation is a state subject, and is looked after by the state governments under the MNP (minimum needs programme) and supplemented by the CRSP (Centrally Sponsored Rural Sanitation Programme) aiming at increased coverage and eradicating manual scavenging. It has been targeted to achieve at least 50% coverage by the end of Ninth Plan period.

Inefficient resource utilisation

With government policies that provide hardly any incentive to encourage efficient usage, water has conventionally been considered as a free commodity. Distribution losses of treated water range between 25% and 40%; losses in irrigation are even to the extent of 45% due to seepage and excess application and storage losses are estimated to be about 15%. Photo 11.2 shows such a case of resource loss due to water overflow. Industrial output per unit of water withdrawal in India is only $5 per cubic metre as against output at $25 and $32 for such developed countries as Japan and Sweden. Even in the domestic sector, areas with high per capita water availability are known for poor water utilisation. The subsidy regime in the domestic sector further leaves the service providers with insufficient funds for proper upkeep of the system and finally affects the quality of service and its efficiency.

State

Resource quantity

Comparing water availability in the country's basins with the standard definition by Engleman and Roy, only four of the twenty basins had more than 1,700 m^3 per capita per year renewable fresh water resources, nine basins had between 1,000 to 1,700 m^3, five basins had between 500 to 1,000 m^3, and two basins had less than 500 m^3 in 1991.

In 1997, at an agro-ecological zone (AEZ) level, availability exceeded requirements throughout the country with the exception of AEZ 2 and AEZ 9 (northern plains) where it is less than the demand. By 2047 it is anticipated that demand will exceed availability in five of the twenty AEZs and of these, the situation will be particularly critical in AEZs 2, 9, and 15 (Bengal and Assam plains) where availability is less than 75% of the total demand.

Despite the fact that the National Water Policy accords top priority to drinking water, the coverage of population for safe water supplies is still less than 100 % in most urban and rural areas. Though maximising water availability, water pricing and water zoning for proper management of resources have all been addressed in this national document, poor execution presents a diametrically opposite picture on ground.

Further, with water being a state subject and with the absence of water policies at the state level, resource management and conflict resolution ultimately affect resource availability. Inadequate legal framework for withdrawals also affects resource accessibility due to overexploitation.

Resource quality

The major rivers of the country have generally retained pristine water quality in the less densely populated upper stretches where the likelihood of getting affected by man's interference is minimal. As the rivers enter the plains, these start getting exploited for irrigation and receiving pollution discharges due to human activities such as intensive agriculture, use of fertilisers and insecticides, domestic sewage, industrial effluents etc.

Thus in the middle stretches, the rivers are most affected both, due to increased water requirement for various consumptive and non-consumptive uses, and degraded water quality. Increased quantity of wastes of a more complex nature finds way into the river and tends to deteriorate the water quality. This makes the situation grave especially during the lean flow season when the amount of dilution water available is less.

The Central Pollution Control has been monitoring water quality of national aquatic resources in collaboration with concerned State Pollution Control Boards at 507 locations, of which 430 stations are under MINARS (Monitoring of Indian National Aquatic Resources), 50 stations are under GEMS (Global Environmental Monitoring Systems) and 27 stations under the YAP (Yamuna Action Plan). The polluted stretches identified in some of the major rivers are based on regular monitoring. The water quality and desired water quality have been classified as A, B, C, D, and E, which reflect the best use of the water.

The water quality monitoring results obtained during 1998 indicate that organic and bacterial contamination still continue to be critical sources of pollution in Indian aquatic resources. The results show that BOD concentrations below 3 mg/l were observed in 61% of the samples as against 65% in 1997. BOD between 3 to 6 mg/ l was observed in 24%

of the samples as against 28% in 1997 and BOD exceeding 6 mg/l was observed in 14% of the samples, which is the same as in 1997. The number of observations having high coliform density increased in 1998 against 1997. The Yamuna river is the most polluted in the country having high BOD and coliform in the stretch between Delhi and Etawah. Other severely polluted rivers are the Sabarmati at Ahmedabad, Gomti at Lucknow, Kali, Adyar, Cooum (entire stretches), Veghai at Madurai, and Musi d/s of Hyderabad.

In an effort to assess the health of a water body, the CPCB is developing a bio-monitoring methodology under the Indo-Dutch Collaboration Programme on the Environment. Intensive exercises have been carried out for the Yamuna River and the methodology has been developed and tested for other rivers. In the first phase, 215 locations have been selected for the introduction of bio-monitoring on the basis of the interpretation of physicochemical data at different locations. CPCB has also carried out limited water quality monitoring of the wells in different states and calculated percent violations over the desired levels of water quality in terms of pH, dissolved oxygen, BOD and total coliform.

Certain locations in these states reported 100% violation over the desired levels for dissolved oxygen and total coliform. CWC's studies on chemical composition of groundwater in phreatic zones have revealed that in many cases anomalously high concentrations of nitrates, potassium and even phosphates are present in contrast to their virtual absence or low concentration (nitrate and potassium < 10 mg/l) in semi-confined and confined aquifers.

The unsystematic use of synthetic fertilisers coupled with improper water management has affected the groundwater quality in many parts of the country. The statewise brief account of the incidence of groundwater pollution also reflects the occurrence of high concentrations of heavy/toxic metals, fluoride and nitrates at different locations around the country. The presence of zinc in shallow aquifers of Delhi is reported at places located close to areas of intensive agricultural practices coupled with extensive use of chemical fertilisers.

Even with strong legislative provisions such as the Water (Prevention and Control of Pollution) Act and the Environment Protection Protection Act, since 1974 and 1986 respectively, 851 defaulting industries were located along the rivers and lakes in 1997. The Water Cess Act, 1977 has also failed to act as a market based instrument in reducing the quantity of polluted discharges.

Response

Existing response

Water has been included in India's Constitution as Entry 17 of the state list. The current institutional arrangement for managing water resources in India involves various government agencies. At the central level, the Ministry of Water Resources (MoWR) is responsible for developing, conserving and managing water as a national resource covering areas as diverse as irrigation, ground water exploitation, drainage and flood control.

The MoWR functions through the Central Water Commission, National Water Development Agency and Central Ground Water Board. The Ministry of Environment and Forests (MoEF) is the nodal agency for water quality and environmental matters. Water supply in urban and rural areas is coordinated by the Ministries of Urban Affairs and Rural Development, respectively. Besides, water is also a subject of several other ministries and departments such as the Ministry of Agriculture, Power, Health and Family welfare, Surface Transport and the Inland Waterways Authority. Major policies and legislations guiding the management of water resources and its quality and initiatives taken have been discussed.

- The National Water Policy, 1987, formulated by the Government of India accords top priority to drinking water supply in the allocation of water resources for various beneficial uses. After drinking water, the list includes irrigation, hydropower, navigation and industrial and other uses. The policy also addresses issues such as planning of water resource development projects, maximising water availability, water pricing, water quality, water zoning for proper management of resources and other issues.
- The government explicitly enacted the Water (Prevention and Control of Pollution) *Act, 1974,* with the primary objective of preventing and controlling of water pollution. The Water Act established the Central Pollution Control Board and the state pollution control boards for its implementation. The Water Act empowers the state pollution control boards to lay down and maintain location and source specific standards for discharge of wastewater. The actual provisions for enforcement such as penalties, imprisonment etc. are confined to source specific standards for individual polluters.
- The Environment Protection Act, 1986, is an umbrella act providing for the protection and improvement of environment and for matters connected therewith. It authorises the central government to intervene directly in order to protect the environment and also allows public interest litigation for the same purpose. The nature of penalties under this act is similar to those authorised under the Water Act.
- The government has also introduced, as a supplementary measure, major economic incentives for pollution abatement, besides the 'command and

control' regulatory mechanism. The *Water Cess Act* was introduced in 1977, empowering the state pollution control boards to levy a cess on local authorities supplying water to consumers and on consumption of water for certain specified activities. The Act also provides for a rebate on the cess payable if the local authority or industry concerned installs a plant to treat sewage or trade effluent. The cess rates were increased three fold in February 1992. A rebate of 25% on the cess payable has been provided to those industries whose wastewater discharge does not exceed the quantity declared by them and which also comply with the effluent standards prescribed under the Water Act and the Environment Protection Act.

— Under the 1994 EIA notification, an *Environmental* Impact Assessment has been made mandatory for 30 categories of development activities involving investments of more than Rs 500 million and above and environmental clearance for activities is given by the MoEF.

— Under the National River Action Plan (NRAP), certain stretches of major rivers with high or intermediate levels of pollution were identified by the CPCB. Sewage collection and treatment works being created to reduce the pollution load to these rivers include schemes for better sewage interception and diversion, construction of sewage treatment plants, provisions for low cost sanitation and other schemes. In the first phase, in the GAP (Ganga action plan), 29 towns were selected along the river and 261 schemes of pollution abatement sanctioned. At present, 156 towns are being considered under the NRAP, out of which about 74 towns are located on the river Ganga, 21 on the river Yamuna, 12 on the Damodar, 6 on the Godavari, 9 on the Cauvery, 4 each on Tungbhadra and Satlej, 3 each on the Subarnarekha, Betwa, Wainganga, Brahmini,

Chambal, Gomti, 2 on the Krishna and one each on the Sabarmati, Khan, Kshipra, Narmada, and Mahanadi.

- To focus on urban lakes subjected to anthropogenic pressures, the National Lake Conservation Plan (NLCP), 1993 was prepared. Bhoj Lake of Madhya Pradesh is already getting assistance under funds provided by OECF, Japan.
- Under the World Bank aided Industrial Pollution Control project there is a provision of loan and grant assistance to proposals of construction of Common Effluent Treatment Plants (CETP) for the treatment of effluents from a cluster of industries particularly of small scale.

Besides, there are other acts that have a bearing on the water policy. The Easement Act, 1882 allows rights to use the groundwater by viewing it as an attachment to land. It also provides that all surface water is to be treated as state property. The Transfer of Property Act, 1994 provides that ownership in groundwater can be given to one only if ownership in land overlying is also transferred. Thus groundwater is viewed essentially as an attachment to land.

Policy gaps

- The major bottleneck in an effective policy formulation and implementation is the current institutional set-up involving various government agencies. Further, there is a separation of responsibilities on the basis of water quality and quantity. As many as eight agencies are involved in collecting data on the following water-related parameters: quality of surface water, ground water quality, monitoring of drinking water quality, sanitation and drinking water supply. Such a fragmentary approach, both at the central and state levels, results in duplication and ambiguity of functions and discourages unitary analysis of this scarce resource. For instance, the CPCB monitors the water quality at 507 locations and the CWC separately measures water quality at another 300 locations. However, co-ordination between the two agencies in fixing the monitoring locations and defining monitoring protocols is missing.

— Water being a state subject, the states are empowered to enact laws or frame policies related to water. Even then, only some of the states have set up organisations for planning and allocating water for various purposes. Though water policy for the country has been prepared by the Ministry of Water Resources (MoWR), only four of the states have their own respective state water policies.

— A proper legal framework for regulating withdrawals of groundwater is not in place. Though efforts have been made to check the overexploitation of groundwater through licensing, credit or electricity restrictions, these restrictions are directed only at the creation of wells. Even the licenses do not monitor or regulate the quantum of water extracted.

— The water cess in industries, is potentially an effective instrument for inducing abatement, but the rates of raw water are so low that the rebate has been as much of an incentive so far. Market-based instruments to encourage resource conservation mainly in the agriculture and domestic sector have not been really tried. This accompanied with the subsidy regime in these sectors has resulted in poor resource usage efficiency.

— It was realised during the later stages of implementation of the Ganga Action Plan, that the local authorities were not able to operate and maintain these assets due to inadequate resources and skills. The level of commitment required from the state agencies was also missing. The pollution arose from a number of diffused sources either urban or rural.

Knowledge/information/data gaps

— Water quality monitoring by CPCB is at present being carried out at 507 locations, as against 77 stations in 1977. The 476 stations at which monitoring was carried out in 1996 and 1997 comprised 407 stations on rivers, 20 on wells, 33 on lakes, 9 on canals, 2 on ponds, 3 on creeks, and 2 on tanks. Water quality monitoring at 383 locations is conducted on a monthly basis, on a quarterly basis at 121 stations and annually for 3 locations through the involvement of several agencies. The maximum, minimum, and mean values of the parameters and the percentage violations for select parameters are reported. However, specific information is not available for water quality in these water bodies for seasons with lean flow. The frequency of monitoring and number of monitoring stations also is not representative of the quality of the water body specifically in the non-monsoon period.

— The CWC also monitors the water quality for 47 parameters at about 300 locations. Though there is a dedicated staff for monitoring the water quality at these three times in a month, published data are not available. Information is available only at the regional office level and that too, on request.

— Information on the availability of groundwater and its quality is limited. Though groundwater availability maps have been prepared for certain locations, extraction rates have not been defined.

— Much of the information—quantitative as well as qualitative, on water supplied, coverage of population, quality of service and sanitation both in the urban and rural areas is not available. Besides, information gaps on water consumption and effluent discharge patterns for industries also exist.

REFERENCES

Gleick PH and associates, *The World's Water: The Biennial Report on Freshwater Resources*, Island Press, Washington, D.C., 1998.

John M. DeMan., *Principles of Food Chemistry 3rd Edition*, 1999.

Jones OA, JN Lester and N Voulvoulis, "Pharmaceuticals: a threat to drinking water?," *Trends in Biotechnology* 23(4): 163, 2005.

Vickie A. Vaclavik and Elizabeth W. Christian., *Essentials of Food Science 2nd Edition*, 2003.

8

Irrigation Systems

The irrigation system consists of a intake structure or pumping station, a conveyance system, a distribution system, a field application system, and a drainage system.

The intake structure, or pumping station, directs water from the source of supply, such as a reservoir or a river, into the irrigation system.

The conveyance system assures the transport of water from the main intake structure or main pumping station up to the field ditches. The distribution system assures the transport of water through field ditches to the irrigated fields. The field application system assures the transport of water within the fields. The drainage system removes the excess water (caused by rainfall and/or irrigation) from the fields.

INTAKE STRUCTURE AND PUMPING STATION

Intake Structure

The intake structure is built at the entry to the irrigation system. (Figure 1) Its purpose is to direct water from the original source of supply (lake, river, reservoir etc.) into the irrigation system.

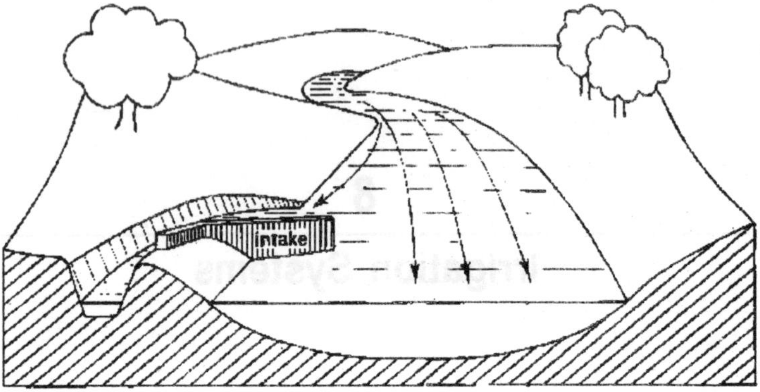

Figure 1. An intake structure

Pumping Station

In some cases, the irrigation water source lies below the level of the irrigated fields. Then a pump must be used to supply water to the irrigation system (Figure 2).

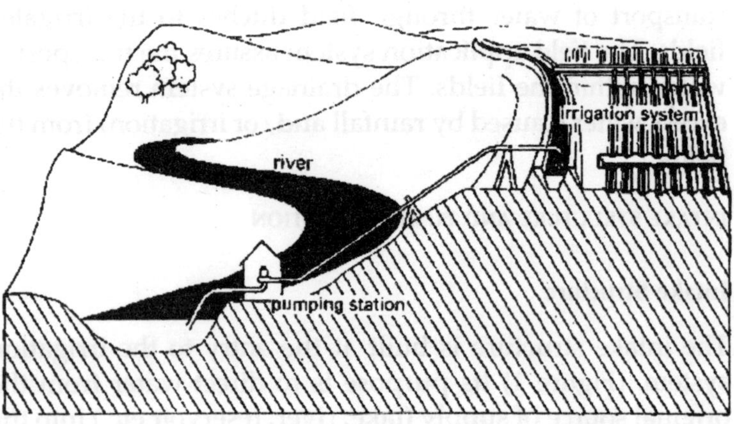

Figure 2. A pumping station

There are several types of pumps, but the most commonly used in irrigation is the centrifugal pump.

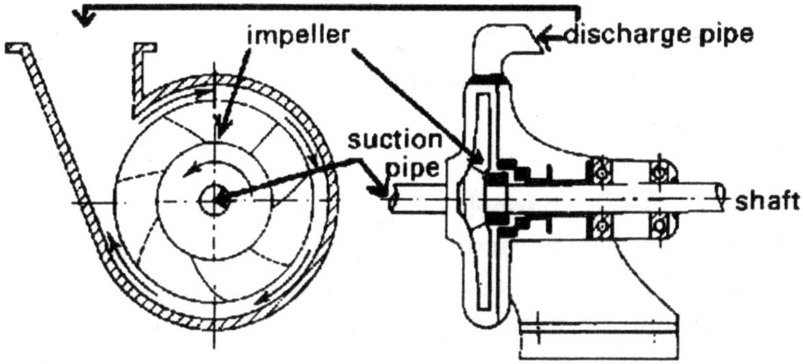

Figure 3a. Diagram of a centrifugal pump

The centrifugal pump (Figure 3a) consists of a case in which an element, called an impeller, rotates driven by a motor (Figure 3b). Water enters the case at the centre, through the suction pipe. The water is immediately caught by the rapidly rotating impeller and expelled through the discharge pipe.

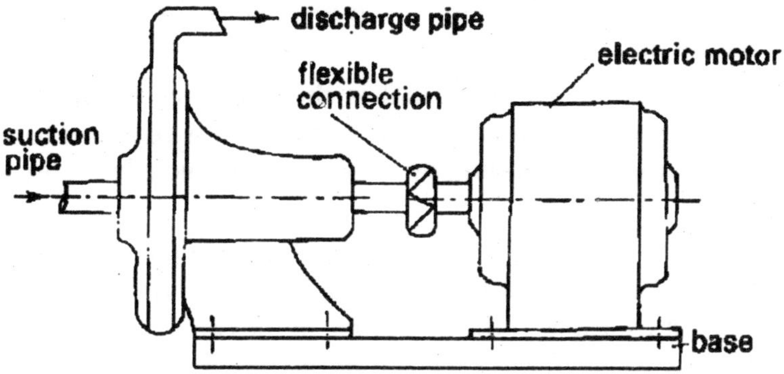

Figure 3b. Centrifugal pump and motor

Conveyance and Distribution System

The conveyance and distribution systems consist of canals transporting the water through the whole irrigation system. Canal structures are required for the control and measurement of the water flow.

Open canals

An open canal, channel, or ditch, is an open waterway whose purpose is to carry water from one place to another. Channels and canals refer to main waterways supplying water to one or more farms. Field ditches have smaller dimensions and convey water from the farm entrance to the irrigated fields.

i) *Canal characteristics* : According to the shape of their cross-section, canals are called rectangular (a), triangular (b), trapezoidal (c), circular (d), parabolic (e), and irregular or natural (f).

The most commonly used canal cross-section in irrigation and drainage, is the trapezoidal cross-section. For the purposes of this publication, only this type of canal will be considered.

The typical cross-section of a trapezoidal canal is shown in Figure 4.

The freeboard of the canal is the height of the bank above the highest water level anticipated. It is required to guard against overtopping by waves or unexpected rises in the water level.

The side slope of the canal is expressed as ratio, namely the vertical distance or height to the horizontal distance or width. For example, if the side slope of the canal has a ratio of 1:2 (one to two), this means that the horizontal distance (w) is two times the vertical distance (h).

The bottom slope of the canal does not appear on the drawing of the cross-section but on the longitudinal section.

It is commonly expressed in percent or per mil.

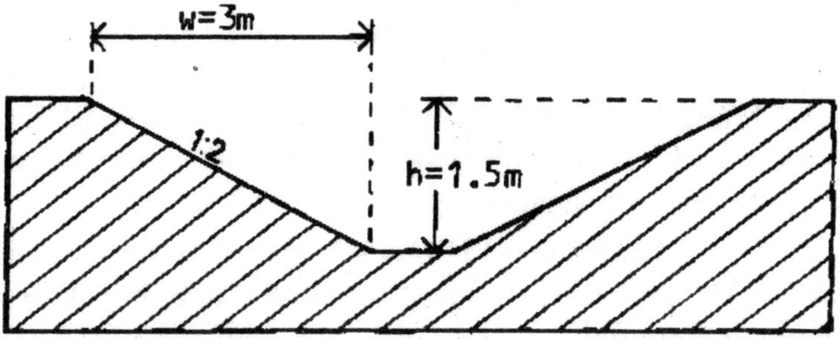

Figure 4. A side slope of 1:2 (one to two)

An example of the calculation of the bottom slope of a canal is given below:

$$\text{The bottom slope (\%)} = \frac{\text{height difference (metres)}}{\text{horizontal distance (metres)}} \times 100 = \frac{1\text{m}}{100\text{m}} \times 100 = 1\%$$

or

$$\text{The bottom slope (\%)} = \frac{\text{height difference (metres)}}{\text{horizontal distance (metres)}} \times 1000 = \frac{1\text{m}}{100\text{m}} \times 1000 = 10\%$$

ii) *Earthen Canals* : Earthen canals are simply dug in the ground and the bank is made up from the removed earth.

The disadvantages of earthen canals are the risk of the side slopes collapsing and the water loss due to seepage. They also require continuous maintenance in order to control weed growth and to repair damage done by livestock and rodents.

iii) *Lined Canals* : Earthen canals can be lined with impermeable materials to prevent excessive seepage and growth of weeds.

Lining canals is also an effective way to control canal bottom and bank erosion. The materials mostly used for canal lining are concrete (in precast slabs or cast in place), brick or rock masonry and asphaltic concrete (a mixture of sand, gravel and asphalt).

The construction cost is much higher than for earthen canals. Maintenance is reduced for lined canals, but skilled labour is required.

Canal structures

The flow of irrigation water in the canals must always be under control. For this purpose, canal structures are required. They help regulate the flow and deliver the correct amount of water to the different branches of the system and onward to the irrigated fields.

There are four main types of structures: erosion control structures, distribution control structures, crossing structures and water measurement structures.

i. Erosion control structures

a. *Canal erosion* : Canal bottom slope and water velocity are closely related, as the following example will show.

A cardboard sheet is lifted on one side 2 cm from the ground. A small ball is placed at the edge of the lifted side of the sheet. It starts rolling downward, following the slope direction. The sheet edge is now lifted 5 cm from the ground, creating a steeper slope. The same ball placed on the top edge of the sheet rolls downward, but this time much faster. The steeper the slope, the higher the velocity of the ball.

Water poured on the top edge of the sheet reacts exactly the same as the ball. It flows downward and the steeper the slope, the higher the velocity of the flow.

Water flowing in steep canals can reach very high velocities. Soil particles along the bottom and banks of an earthen canal are then lifted, carried away by the water flow, and deposited downstream where they may block the canal and silt up structures. The canal is said to be under erosion; the banks might eventually collapse.

b. *Drop structures and chutes* :Drop structures or chutes are required to reduce the bottom slope of canals lying on steeply sloping land in order to avoid high velocity of the flow and risk of erosion. These structures permit the canal to be constructed as a series of relatively flat sections, each at a different elevation. (Figure 5)

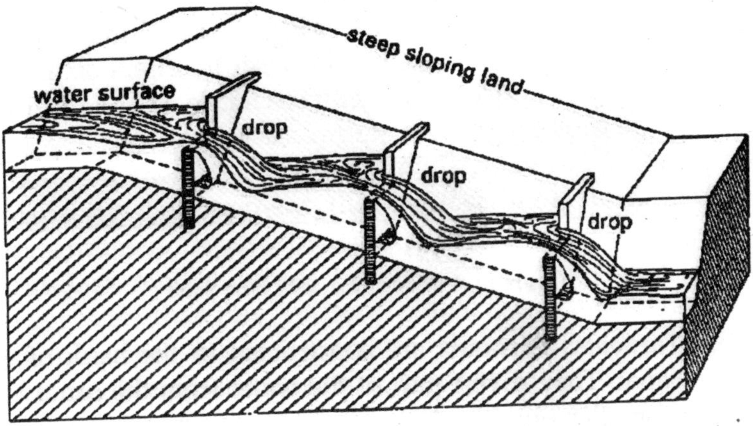

Figure 5. Longitudinal section of a series of drop structures

Drop structures take the water abruptly from a higher section of the canal to a lower one. In a chute, the water does not drop freely but is carried through a steep, lined canal section. Chutes are used where there are big differences in the elevation of the canal.

ii. Distribution control structures

Distribution control structures are required for easy and accurate water distribution within the irrigation system and on the farm.

a. *Division boxes* :Division boxes are used to divide or direct the flow of water between two or more canals or ditches. Water enters the box through an opening on one side and flows out through openings on the other sides. These openings are equipped with gates (Figure 6).

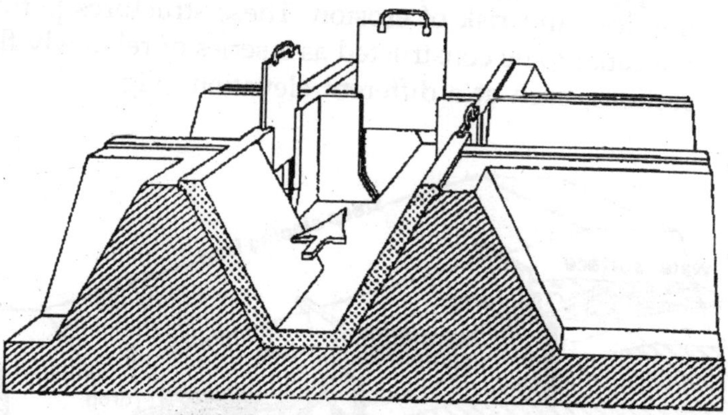

Figure 6. A division box with three gates

b. *Turnouts:* Turnouts are constructed in the bank of a canal. They divert part of the water from the canal to a smaller one.

Turnouts can be concrete structures, or pipe structures.

c. *Checks* : To divert water from the field ditch to the field, it is often necessary to raise the water level in the ditch. Checks are structures placed across the ditch to block it temporarily and to raise the upstream water level. Checks can be permanent structures or portable.

iii) Crossing structures

It is often necessary to carry irrigation water across roads, hillsides and natural depressions. Crossing structures, such as flumes, culverts and inverted siphons, are then required.

a) *Flumes* : Flumes are used to carry irrigation water across gullies, ravines or other natural depressions. They are open canals made of wood (bamboo), metal or concrete which often need to be supported by pillars.

b) *Culverts* : Culverts are used to carry the water across roads. The structure consists of masonry or concrete headwalls at the inlet and outlet connected by a buried pipeline.

c. *Inverted siphons* : When water has to be carried across a road which is at the same level as or below the canal bottom, an inverted siphon is used instead of a culvert. The structure consists of an inlet and outlet connected by a pipeline. Inverted siphons are also used to carry water across wide depressions.

iv) Water measurement structures

The principal objective of measuring irrigation water is to permit efficient distribution and application. By measuring the flow of water, a farmer knows how much water is applied during each irrigation.

In irrigation schemes where water costs are charged to the farmer, water measurement provides a basis for estimating water charges.

The most commonly used water measuring structures are weirs and flumes. In these structures, the water depth is read on a scale which is part of the structure. Using this reading, the flow-rate is then computed from standard formulas or obtained from standard tables prepared specially for the structure.

a) *Weirs* : In its simplest form, a weir consists of a wall of timber, metal or concrete with an opening with fixed dimensions cut in its edge. The opening, called a notch, may be rectangular, trapezoidal or triangular.

b) *Parshall flumes :* The Parshall flume consists of a metal or concrete channel structure with three main sections: (1) a converging section at the upstream end, leading to (2) a constricted or throat section and (3) a diverging section at the downstream end.

Depending on the flow condition (free flow or submerged flow), the water depth readings are taken on one scale only (the upstream one) or on both scales simultaneously.

c) *Cut-throat flume:* The cut-throat flume is similar to the Parshall flume, but has no throat section, only converging and diverging sections. Unlike the Parshall flume, the cut-throat flume has a flat bottom. Because it is easier to construct and install, the cut-throat flume is often preferred to the Parshall flume.

Field Application Systems

There are many methods of applying water to the field. The simplest one consists of bringing water from the source of supply, such as a well, to each plant with a bucket or a water-can.

This is a very time-consuming method and it involves quite heavy work. However, it can be used successfully to irrigate small plots of land, such as vegetable gardens, that are in the neighbourhood of a water source.

More sophisticated methods of water application are used in larger irrigation systems. There are three basic methods: surface irrigation, sprinkler irrigation and drip irrigation.

Surface irrigation

Surface irrigation is the application of water to the fields at ground level. Either the entire field is flooded or the water is directed into furrows or borders.

i) Furrow irrigation

Furrows are narrow ditches dug on the field between the rows of crops. The water runs along them as it moves down the slope of the field.

The water flows from the field ditch into the furrows by opening up the bank or dyke of the ditch or by means of syphons or spiles. Siphons are small curved pipes that deliver water over the ditch bank. Spiles are small pipes buried in the ditch bank.

ii) Border irrigation

In border irrigation, the field to be irrigated is divided into strips (also called borders or borderstrips) by parallel dykes or border ridges.

The water is released from the field ditch onto the border through gate structures called outlets. The water can also be released by means of siphons or spiles. The sheet of flowing water moves down the slope of the border, guided by the border ridges.

iii) Basin irrigation

Basins are horizontal, flat plots of land, surrounded by small dykes or bunds. The banks prevent the water from flowing to the surrounding fields. Basin irrigation is commonly used for rice grown on flat lands or in terraces on hillsides. Trees can also be grown in basins, where one tree usually is located in the centre of a small basin.

Sprinkler irrigation

With sprinkler irrigation, artificial rainfall is created. The water is led to the field through a pipe system in which the water is under pressure. The spraying is accomplished by using several rotating sprinkler heads or spray nozzles or a single gun type sprinkler.

Drip irrigation

In drip irrigation, also called trickle irrigation, the water is led to the field through a pipe system. On the field, next to the row of plants or trees, a tube is installed. At regular intervals, near the plants or trees, a hole is made in the tube and equipped with an emitter. The water is supplied slowly, drop by drop, to the plants through these emitters.

Drainage System

A drainage system is necessary to remove excess water from the irrigated land. This excess water may be e.g. waste water from irrigation or surface runoff from rainfall. It may also include leakage or seepage water from the distribution system.

Excess surface water is removed through shallow open drains. Excess groundwater is removed through deep open drains or underground pipes.

Need for drainage

During rain or irrigation, the fields become wet. The water infiltrates into the soil and is stored in its pores. When all the pores are filled with water, the soil is said to be saturated and no more water can be absorbed; when rain or irrigation continues, pools may form on the soil surface.

Part of the water present in the saturated upper soil layers flows downward into deeper layers and is replaced by water infiltrating from the surface pools.

When there is no more water left on the soil surface, the downward flow continues for a while and air re-enters in the pores of the soil. This soil is not saturated anymore.

However, saturation may have lasted too long for the plants' health. Plant roots require air as well as water and most plants cannot withstand saturated soil for long periods (rice is an exception).

Besides damage to the crop, a very wet soil makes the use of machinery difficult, if not impossible.

The water flowing from the saturated soil downward to deeper layers, feeds the groundwater reservoir. As a result, the groundwater level (often called groundwater table or simply water table) rises. Following heavy rainfall or continuous over-irrigation, the groundwater table may even reach and saturate part of the rootzone. Again, if this situation lasts too long, the plants may suffer. Measures to control the rise of the water table are thus necessary.

The removal of excess water either from the ground surface or from the rootzone, is called drainage.

Excess water may be caused by rainfall or by using too much irrigation water, but may also have other origins such as canal seepage or floods.

In very dry areas there is often accumulation of salts in the soil. Most crops do not grow well on salty soil. Salts can be washed out by percolating irrigation water through the rootzone of the crops. To achieve sufficient percolation, farmers will apply more water to the field than the crops need. But the salty percolation water will cause the water table to rise. Drainage to control the water table, therefore, also serves to control the salinity of the soil.

Different Types of Drainage

Drainage can be either natural or artificial. Many areas have some natural drainage; this means that excess water flows from the farmers' fields to swamps or to lakes and rivers. Natural drainage, however, is often inadequate and artificial or man-made drainage is required. There are two types of artificial drainage: surface drainage and subsurface drainage.

Surface drainage

Surface drainage is the removal of excess water from the surface of the land. This is normally accomplished by shallow ditches, also called open drains. The shallow ditches discharge into larger and deeper collector drains. In order to facilitate the flow of excess water toward the drains, the field is given an artificial slope by means of land grading.

Subsurface drainage

Subsurface drainage is the removal of water from the rootzone. It is accomplished by deep open drains or buried pipe drains.

i) Deep open drains

The excess water from the rootzone flows into the open drains. The disadvantage of this type of subsurface drainage is that it makes the use of machinery difficult.

ii) Pipe drains

Pipe drains are buried pipes with openings through which the soil water can enter. The pipes convey the water to a collector drain.

Drain pipes are made of clay, concrete or plastic. They are usually placed in trenches by machines. In clay and concrete pipes (usually 30 cm long and 5 - 10 cm in diameter) drainage water enters the pipes through the joints. Flexible plastic drains are much longer (up to 200 m) and the water enters through perforations distributed over the entire length of the pipe.

iii) Deep open drains versus pipe drains

Open drains use land that otherwise could be used for crops. They restrict the use of machines. They also require a large number of bridges and culverts for road crossings and access to the fields. Open drains require frequent maintenance (weed control, repairs, etc.).

In contrast to open drains, buried pipes cause no loss of cultivable land and maintenance requirements are very limited. The installation costs, however, of pipe drains may be higher due to the materials, the equipment and the skilled manpower involved.

CHOOSING AN IRRIGATION METHOD

To choose an irrigation method, the farmer must know the advantages and disadvantages of the various methods. He or she must know which method suits the local conditions best. Unfortunately, in many cases there is no single best solution: all methods have their advantages and disadvantages. Testing of the various methods—under the prevailing local conditions—provides the best basis for a sound choice of irrigation method.

Surface, Sprinkler or Drip Irrigation

The suitability of the various irrigation methods, i.e. surface, sprinkler or drip irrigation, depends mainly on the following factors:

- natural conditions
- type of crop
- type of technology
- previous experience with irrigation
- required labour inputs
- costs and benefits.

NATURAL CONDITIONS

The natural conditions such as soil type, slope, climate, water quality and availability, have the following impact on the choice of an irrigation method:

Soil type:	Sandy soils have a low water storage capacity and a high infiltration rate. They therefore need frequent but small irrigation applications, in particular when the sandy soil is also shallow. Under these circumstances, sprinkler or drip irrigation are more suitable than surface irrigation. On loam or clay soils all three irrigation methods can be used, but surface irrigation is more commonly found. Clay soils with low infiltration rates are ideally suited to surface irrigation. When a variety of different soil types is found within one irrigation scheme, sprinkler or drip irrigation are recommended as they will ensure a more even water distribution.
Slope:	Sprinkler or drip irrigation are preferred above surface irrigation on steeper or unevenly sloping lands as they require little or no land levelling. An exception is rice grown on terraces on sloping lands.
Climate:	Strong wind can disturb the spraying of water from sprinklers. Under very windy conditions, drip or surface irrigation methods are preferred. In areas of supplementary irrigation, sprinkler or drip irrigation may be more suitable than surface irrigation because of their flexibility and adaptability to varying irrigation demands on the farm.
Water availability:	Water application efficiency is generally higher with sprinkler and drip irrigation than surface irrigation and so these methods are preferred when water is in short supply. However, it must be remembered that efficiency is just as much a function of the irrigator as the method used.
Water quality:	Surface irrigation is preferred if the irrigation water contains much sediment. The sediments may clog the drip or sprinkler irrigation systems. If the irrigation water contains dissolved salts, drip irrigation is particularly suitable, as less water is applied to the soil than with surface methods. Sprinkler systems are more efficient that surface irrigation methods in leaching out salts.

TYPE OF CROP

Surface irrigation can be used for all types of crops.

Sprinkler and drip irrigation, because of their high capital investment per hectare, are mostly used for high value cash crops, such as vegetables and fruit trees. They are seldom used for the lower value staple crops.

Drip irrigation is suited to irrigating individual plants or trees or row crops such as vegetables and sugarcane. It is not suitable for close growing crops.

TYPE OF TECHNOLOGY

The type of technology affects the choice of irrigation method. In general, drip and sprinkler irrigation are technically more complicated methods. The purchase of equipment requires high capital investment per hectare. To maintain the equipment a high level of 'know-how' has to be available,. Also, a regular supply of fuel and spare parts must be maintained which—together with the purchase of equipment—may require foreign currency.

Surface irrigation systems—in particular small-scale schemes—usually require less sophisticated equipment for both construction and maintenance (unless pumps are used). The equipment needed is often easier to maintain and less dependent on the availability of foreign currency.

PREVIOUS EXPERIENCE WITH IRRIGATION

The choice of an irrigation method also depends on the irrigation tradition within the region or country. Introducing a previously unknown method may lead to unexpected complications. It is not certain that the farmers will accept the new method. The servicing of the equipment may be problematic and the costs may be high compared to the benefits.

Often it will be easier to improve the traditional irrigation method than to introduce a totally new method.

REQUIRED LABOUR INPUTS

Surface irrigation often requires a much higher labour

input—for construction, operation and maintenance—than sprinkler or drip irrigation. Surface irrigation requires accurate land levelling, regular maintenance and a high level of farmers' organisation to operate the system. Sprinkler and drip irrigation require little land levelling; system operation and maintenance are less labour-intensive.

COSTS AND BENEFITS

Before choosing an irrigation method, an estimate must be made of the costs and benefits of the available options. On the cost side not only the construction and installation, but also the operation and maintenance (per hectare) should be taken into account. These costs should then be compared with the expected benefits (yields). It is obvious that farmers will only be interested in implementing a certain method if they consider this economically attractive. Cost/benefit analysis is, however, beyond the scope of this manual.

Basin, Furrow or Border Irrigation

This section discusses some of the important factors which should be taken into account when determining which surface irrigation method is most suitable: basin, furrow or border irrigation. Again, it is not possible to give specific guidelines leading to a single best solution; each option has its advantages and disadvantages. Factors to be taken into account include:

— natural circumstances (slope, soil type)
— type of crop
— required depth of irrigation application
— level of technology
— previous experience with irrigation
— required labour inputs.

NATURAL CIRCUMSTANCES

Flat lands, with a slope of 0.1% or less, are best suited for basin irrigation: little land levelling will be required. If the slope is more than 1%, terraces can be constructed. However, the amount of land levelling can be considerable.

Furrow irrigation can be used on flat land (short, near horizontal furrows), and on mildly sloping land with a slope of maximum 0.5%. On steeper sloping land, contour furrows can be used up to a maximum land slope of 3%. A minimum slope of 0.05% is recommended to assist drainage.

Border irrigation can be used on sloping land up to 2% on sandy soil and 5% on clay soil. A minimum slope of 0.05% is recommended to ensure adequate drainage.

Surface irrigation may be difficult to use on irregular slopes as considerable land levelling may be required to achieve the required land gradients.

All soil types, except coarse sand with an infiltration rate of more than 30 mm/hour, can be used for surface irrigation. If the infiltration rate is higher than 30 mm/hour, sprinkler or drip irrigation should be used.

TYPE OF CROP

Paddy rice is always grown in basins. Many other crops can also be grown in basins: e.g. maize, sorghum, trees, etc. Those crops that cannot stand a very wet soil for more than 12-24 hours should not be grown in basins.

Furrow irrigation is best used for irrigating row crops such as maize, vegetables and trees.

Border irrigation is particularly suitable for close growing crops such as alfalfa, but border irrigation can also be used for row crops and trees.

REOUIRED DEPTH OF IRRIGATION APPLICATION

When the irrigation schedule has been determined it is

known how much water (in mm) has to be given per irrigation application. It must be checked that this amount can indeed be given, with the irrigation method under consideration.

Field experience has shown that most water can be applied per irrigation application when using basin irrigation, less with border irrigation and least with furrow irrigation. In practice, in small-scale irrigation projects, usually 40-70 mm of water are applied in basin irrigation, 30-60 mm in border irrigation and 20-50 mm in furrow irrigation.

This means that if only little water is to be applied per application, e.g. on sandy soils and a shallow rooting crop, furrow irrigation would be most appropriate. (However, none of the surface irrigation methods can be used if the sand is very coarse, i.e. if the infiltration rate is more than 30 mm/hour.)

If, on the other hand, a large amount of irrigation water is to be applied per application, e.g. on a clay soil and with a deep rooting crop, border or basin irrigation would be more appropriate.

The above considerations have been summarised in Table 1. The net irrigation application values used are only a rough guide. They result from a combination of soil type and rooting depth. For example: if the soil is sandy and the rooting depth of the crop is medium, it is estimated that the net depth of each irrigation application will be in the order of 35 mm. The last column indicates which irrigation method is most suitable. In this case medium furrows or short borders.

Level of Technology

Basin irrigation is the simplest of the surface irrigation methods. Especially if the basins are small, they can be constructed by hand or animal traction. Their operation and maintenance is simple.

Furrow irrigation—with the possible exception of short, level furrows -requires accurate field grading. This is often done by machines. The maintenance—ploughing and furrowing—is also often done by machines. This requires skill, organisation and frequently the use of foreign currency for fuel, equipment and spare parts.

Table 1: Selection of an irrigation method based on the depth of the net irrigation application

Soil type	*Rooting depth of the crop*	*Net irrigation depth per application (mm)*	*Irrigation method*
Sand	shallow	20-30	short furrows
	medium	30-40	medium furrows, short borders
	deep	40-50	long furrows, medium borders, small basins
Loam	shallow	30-40	medium furrows, short borders
	medium	40-50	long furrows, medium borders, small basins
	deep	50-60	long borders, medium basins
Clay	shallow	40-50	long furrows, medium borders, small basins
	medium	50-60	long borders, medium basins
	deep	60-70	large basins

Short, level furrows—also called furrow basins—can, like basins, be constructed and maintained by hand. Borders require the highest level of sophistication. They are constructed and maintained by machines. The grading needs to be accurate. Machine operation requires a high level of skill, organisation and usually foreign currency.

PREVIOUS EXPERIENCE WITH IRRIGATION

If there is no tradition in irrigation, the most simple

irrigation method to introduce is basin irrigation. The smaller the basins, the easier their construction, operation and maintenance.

If irrigation is used traditionally, it is usually simpler to improve the traditional irrigation method than it is to introduce a previously unknown method.

REQUIRED LABOUR INPUTS

The required labour inputs for construction and maintenance depend heavily on the extent to which machinery is used. In general it can be stated that to operate the system, basin irrigation requires the least labour and the least skill. For the operation of furrow and border irrigation systems more labour is required combined with more skill.

REFERENCES

Adams, R. M., *Heartland of Cities, Surveys of Ancient Settlement and Land Use on the Central Floodplain of the Euphrates*, Chicago, IL: University of Chicago Press, 1965.

Biswas, A. K., *History of Hydrology*, Amsterdam, Netherlands: North-Holland, 1970.

Butzer, K. W., *Early Hydraulic Civilization in Egypt*, Chicago, IL: University of Chicago Press, 1976.

Doolittle, W. E., *Canal Irrigation in Prehistoric Mexico*, Austin: University of Texas Press, 1990.

Neugebauer, O., *The Exact Sciences in Antiquity*, 2nd ed. New York: Barnes and Noble, 1993.

Vivian, R. G. "Conservation and Diversion: Water-Control Systems in the Anasazi >Southwest.," In *Irrigation's Impact on Society*, Theodore Downing and McGuire Gibson (ed.)., Anthropological Papers of the University of Arizona 25. Tucson: 1974.

Wittfogel, K. A., *The Hydraulic Civilization: Man's Role in Changing the Earth*, Chicago, IL: University of Chicago Press, 1956.

9

Drainage Water Management

In many irrigated areas around the world, rising water tables have subsequently led to waterlogging and associated salinity problems. This has happened where drainage development has not kept pace with irrigation development or where maintenance of drainage facilities has largely been neglected. In response to the increasing world population and economic growth, water withdrawals for human consumption will increase, so increasing the competition for water between municipal, industrial, agricultural, environmental and recreational needs. If present trends continue with water withdrawal under present practices and policies, it is estimated that by 2025 water stress will increase in more than 60 percent of the world.

In this respect, providing food for the growing population is a major challenge as agriculture is already by far the largest water consumer in most regions in the world, except North America and Europe. On a global basis, agriculture accounts for 69 percent of all water withdrawals. Although the water resources for agriculture are often overused and misused, the general belief is that irrigated agriculture has to expand by 20-30 percent in area by 2025 in order to produce sufficient food for the growing world population. In order to avoid exacerbating the water crisis and to prevent considerable food shortages, the productivity of water use needs to increase.

In other words, the amount of food produced with the same amount of water needs to increase. This is possible through the conservation and reuse of the available water resources in the agriculture sector, including usable drainage waters. The overuse and misuse of water in irrigated agriculture has not only resulted in large-scale waterlogging and salinity and overexploitation of groundwater resources, but also in the depriving of downstream users of sufficient water and in the pollution of fresh water resources with contaminated irrigation return flow and deep percolation losses. Water pollution adds to the competition for scarce water resources as it makes them less suitable for other potential beneficial downstream uses. Furthermore, it might cause severe environmental pollution and threaten public health.

Until ten years ago drainage water management received little attention. Drainage research tended to focus on design issues, while evaluation dealt largely with the performance of the installed system in relation to the design criteria. After the 1992 Earth Summit, the international irrigation and drainage community focused its full attention on drainage water management.

Agenda 21 not only stresses the need for drainage as a necessary complement to irrigation development in arid and semi-arid areas, but at the same time it urges the conservation and recycling of freshwater resources in a context of integrated resource management. In many countries these concerns and especially the concern for water quality degradation have resulted in drainage water disposal regulations to maintain the water quality standards of freshwater bodies for other uses, i.e. agricultural, municipal, industrial, environmental and recreational uses.

CHANGE IN DRAINAGE WATER MANAGEMENT

When planners, decision-makers or engineers face the need

for a change in drainage water management, the nature of the exact problem is often not clear and the perceived problem depends considerably on the individual's viewpoint. For example, a farmer's perception of a problem related to drainage water management depends on the physical conditions within the farm boundaries and may differ substantially from those of an irrigation district, national water resource authority or environmental pressure group. In such situations, a soft system approach can help define the problem and seek solutions. A key feature of the soft system approach is that it attempts to avoid identifying problems and seeking solutions from only one perspective and excluding others.

The characteristic of a system is that, although it can be divided into subsystems, it functions as a whole to achieve its objectives. The successful functioning of a system depends on how well it satisfies changing external and internal demands. The system itself is part of a broader universe. In all natural or agricultural systems, there exists a hierarchy of levels. For irrigated crop production this hierarchy might be:

- biochemical and physical systems;
- plant and cropping system;
- farming system;
- irrigation and drainage systems;
- regional, river- or drainage-basin system; and
- supra-regional systems.

It shows some main characteristics of a system as described by the Open University.

1. Defining the boundaries of a system is not a simple task. For example, there could be questions as to whether it would be better to include the regional drainage system within the system boundaries and include pressure from environmental groups, other

water users and water quality rules and regulations in the system's environment. The boundaries chosen reflect the perception of the systems analysts and their understanding of the system's behaviour, which may not always coincide with an organisational or departmental unit. System boundaries in drainage water management might more often coincide with hydrological units.

2. The elements included in the system's environment are those which influence the system in an important way but over which the system itself has no control because they are driven by forces external to the system of interest.
3. The elements within the system are functional working parts of the system. The way that the system operates and behaves depends on the interactions between the elements in the system. Elements could be decomposed into smaller subsystems. The level of detail depends on the specific objectives of the systems analysis. For example, in systems analysis for drainage water management, it would be important to show individual farms as subsystems because on-farm water management defines to a great extent the total quantity and quality of the generated drainage water.
4. As the creator of the system image, a systems analyst defines a particular perspective while another analyst of differing disciplinary training may produce a different image.

When applying a systems approach, at first it is necessary to include all elements in the picture whether they relate to physical, technical, economic, legal, political or administrative considerations as well as any subjective considerations based on the understanding, norms, values and beliefs of the stakeholders involved. A premature exclusion of important elements may result in a suboptimal course of action.

Where changes in drainage water management are necessary, a simple method to produce predetermined results might not work. Rather, a more open-ended investigative approach is required in which important avenues are explored and considered and in which there is room for iterative processes.

The outcome of such an investigative approach is not necessarily the optimal final solution but rather a solution that seems best for those involved, for that particular time and under those particular circumstances. As the external environment and internal demands change constantly, the mix of solutions and needs for change are changing continuously. Moreover, the achieved changes themselves give new insights into processes and enable a continuous learning process.The soft system methodology is a seven-stage approach that has been adopted and adjusted for numerous purposes.

The soft system methodology was developed for managing changes in the context of human activity systems, i.e. a set of activities undertaken by people linked together in a logical structure to constitute a purposeful whole. As drainage water management is a complex mix of human activity and natural and designed systems. Rather it is a free interpretation to illustrate the steps that in general need to be taken to identify the actual problems and search for possible courses of action.

Defining the Problem

Defining the problem and diagnosing the causes is the key to seeking solutions. In the context of drainage water management a problem might be defined as: a need, request or desire for a change in present situation subject to a number of conditions or criteria that must be satisfied simultaneously. This definition implies that a problem situation does not exist independent of the stakeholders who perceive the problem. The solutions to the problem are subject to criteria and conditions based on people's perception of the ideal situation. The conditions and criteria are subjective insofar as they are based on personal or communal objectives influenced by local to national economic, social, cultural, ecological and legal motives, norms and values.

Step 1. An analysis of the problem and the search for solutions starts in a situation where someone or a group of people perceives that there is a problem. At this stage, it might not be possible to define the problem with precision, as different people involved will have differing perceptions of the problem.

Step 2. The first step in formulating the problem more precisely is to identify all the stakeholders involved and their relation to the problem. To form what is called the rich picture, all the elements must be included whether they relate to physical, technical, economic, legal, political or administrative considerations along with subjective considerations based on understanding, norms, values and beliefs of the stakeholders involved. It is then necessary to extract areas of conflict or disagreement as well as the key tasks that must be undertaken within the problem situation.

Step 3. Once the problem situation has been analysed and expressed from the key issues, the relevant systems and subsystems can be defined. These systems can be formal or informal and are those that carry out purposeful activities that will lead to improvement or elimination of the problem situation. Examples of a relevant system within the context of drainage water management might be an on-farm water management system, a system for development of water quality rules and regulation for drainage water disposal, or a land-use planning system. An analysis of the various elements involved provides valuable insight into different perspectives on and constraints surrounding the situation.

For each relevant system, a root definition can be formulated.

A root definition is a formulation of the relevant system and the purpose of the system to achieve a situation in which the problem is balanced out or eliminated. Each root definition provides a particular perspective of the system under investigation. In general, a root definition should include the following information: what the purposeful activity carried out by the system is; who the 'owner' of the system is; who the beneficiaries/victims of the purposeful activity are; who will implement the activity; and what the constraints in its environment are that surround the system. A root definition for on-farm water management might be:

> In an on-farm water management system the responsible farmer uses irrigation water in such a manner that the drainage water generated is of such quantity and quality as permitted by the drainage disposal act whilst maintaining long-term favourable soil conditions that guarantee the production of valuable crops to ensure the financial sustainability of the farming enterprise.

A root definition for the development of water quality rules and regulation could be:

> A governmental system in which water quality rules and regulations for the disposal of agricultural drainage water are promulgated such that they will guarantee water quality for beneficial downstream water uses, including the maintenance of valuable ecosystems, while ensuring the economic sustainability of the agriculture sector.

Spatial Issues

Drainage water problems vary in space and time due mainly to soil heterogeneity and water management practices. The following example illustrates how spatial variability needs to be taken into account in the problem analysis, and also how it influences the options for drainage water management.

Environmental problems related to agricultural drainage water disposal on the western side of the San Joaquin Valley, California, the United States of America, have created a need for improved irrigation water and salt management. The presence of harmful trace elements, mainly selenium, in the drainage water is of major concern and has led to limitations on drainage water disposal to rivers and impoundments such as agricultural evaporation basins. Where water districts in the problem area fail to meet selenium and salt load targets, they risk monetary penalties and loss of access to disposal sites.

Young and Wallender raised the question of whether the constraints raised by current or future regulations will reduce drainage to the point where salt accumulation would occur and in which areas this might occur first. Second, they raised the question of the spatial distribution of drainage water disposal costing strategies. To answer these questions, they developed a methodology for the Panoche Water District to calculate the spatial distribution of water-, salt- and selenium-balance components using data collected by the water district. Furthermore, they developed and evaluated spatially distributed drainage water disposal costing strategies.

Definition of Drainage Water Management and Tasks Involved

In the context of this publication, drainage water management refers to the management and control over the quantity and quality of the drainage water generated in an agricultural drainage basin in arid and semi-arid areas and its final safe disposal. This is achieved through irrigation water conservation measures and the reuse, disposal and treatment of drainage water. Managing drainage water at the field, irrigation-scheme and river-basin levels entails a number of activities including:

— regulating water table levels in the drainage system to ensure the maintenance of favourable soil moisture conditions for optimal crop growth and salinity control;

- developing irrigation and drainage water management strategies to ensure that disposal regulations and water quality standards are met and dealing with issues, problems and conflicts that might occur;
- setting distribution priorities and criteria for reuse in water scarce areas; and
- establishing cost sharing imposed on stakeholders for the use of poor quality water and required treatment to meet the water quality standards for drain discharge.

Driving Forces Behind Drainage Water Management

The main reasons for developing a drainage water management strategy are: (i) prevention of economic and agricultural losses from waterlogging, salinisation and water quality degradation; (ii) concern for quality degradation of shared water resources; and (iii) the need to conserve water for different water users under conditions of actual or projected water scarcity. In addition, the need to comply with drainage water policies and regulations can provide a strong incentive for improved drainage water management.

Many countries and states, e.g. Australia, India, Egypt and California, the United States of America, have a drainage water disposal policy and drainage effluent disposal regulations. In California, the United States of America, and in Australia, the drainage policy guidelines consist of difficult but achievable targets with an active enforcement of regulations. India and Egypt have policy guidelines of a general nature but have not reached the maturity of Californian and Australian laws. Law enforcement often fails in these countries, mainly as a result of administrative shortcomings and unrealistic quality guidelines for their conditions and resources. In these countries and in countries where clear laws and regulations are absent, the prevention of economic and agricultural losses and the conservation of water for other beneficial uses will normally be the main driving factors behind the development of a drainage water management strategy.

Conservation Measures

A major goal of conservation measures is to reduce the volume of drainage effluent generated and the mass discharge of salts and other constituents of concern while at the same time saving water for other beneficial uses. Conservation measures can directly affect the need for and extent of reuse as well as the quantity and quality of drainage effluent requiring disposal and/or treatment.

Where competition for water quantity and quality among different groups of users is a major issue and where drainage effluent disposal is constrained (as in a closed drainage basin) or threatens ecologically sensitive areas, conservation measures are among the first to be considered. The various conservation measures include: source reduction; shallow groundwater table management; groundwater management; and land retirement.

The first line of action in on-farm water conservation is source reduction or reducing deep percolation. Where the goal is not achievable by source reduction, it may be necessary to implement other conservation measures in combination with reuse measures. As different crops require different amounts of water for optimal crop growth and as rooting patterns differ among crops, source reduction and water table management could be combined with a change in cropping pattern to optimise the desired effect of the measures. Changes in cropping pattern should only be considered where compatible with the broader development objectives for the area under consideration.

Reuse Measures

The major aim of reuse measures is to reduce the amount of drainage effluent while at the same time making additional water available for irrigation and other purposes. Reuse measures comprise: reuse in conventional agriculture; reuse in saline agriculture; Integrated Farm Drainage Management (IFDM) systems; reuse in wildlife habitats, wetlands and pastures; and reuse for initial reclamation of salt-affected lands.

Reuse measures can be implemented in combination with conservation measures. Where the drainage water is of relatively good quality, its reuse potential in conventional agriculture is high. Where it is moderately to highly saline, its reuse may be limited to salt tolerant plants. In California, the United States of America, a new IFDM system reuses drainage water sequentially in high water table lands with no opportunities for disposal of subsurface drainage.

Freshwater is used to grow salt sensitive crops, and subsurface drainage water from it is reused to grow salt tolerant crops. Drainage water from the salt tolerant cropland is used to irrigate salt tolerant grasses and halophytes. When the drainage water is no longer usable, it is disposed into solar evaporators for salt harvest. In this system and others, it is always necessary to generate a minimum volume of drainage effluent to prevent the rootzone from becoming too contaminated for any beneficial water use activity. Similar systems are being developed and tested in Australia.

Treatment Measures

Drainage water treatment in a drainage water management plan normally takes place only under severe constraints, such as stringent regulations on disposal of saline drainage waters into streams, or severe water shortage. The drainage water treatment options are based on physical, chemical and/or biological processes. Many of these processes are borrowed from water treatment processes for drinking-water, sewage and industrial wastewater. A few are new processes to remove selenium. The water quality requirements of the treated water need to be well understood prior to the selection of any treatment measure.

Treatment measures such as reverse osmosis are normally only considered for high-value purposes such as drinking-water supply. One option is to partially treat saline waters to a level that could be used for agriculture (e.g. desalt to about 1 dS/m instead of less than 0.1 dS/m). However, the cost of brine management and disposal remains a major problem.

An exception to the typical high cost is the treatment of drainage water through constructed wetlands. Investigations are underway in California, the United States of America, in which wetland cells are planted with cattails, saltmarsh bulrush, tule, baltic rush, saltgrass, smooth cord grass and rabbitsfoot grass. The cells remove about 60-80 percent of the mass of selenium with hydraulic residence times ranging from 7 to 21 days. This treated water is then disposed into agricultural evaporation ponds with reduced toxic selenium effects on waterbirds. Constructed wetlands appear to have some potential to protect aquatic ecosystems and fisheries either downstream or in closed basins.

Disposal Measures

Even after the successful implementation of conservation and reuse measures, there will always be a residual volume of drainage effluent requiring disposal. Disposal options depend mainly on the situation of the drainage outlet in relation to natural disposal sites such as rivers, streams, lakes and oceans. Disposal options to surface water bodies comprise discharge into rivers and streams, river mouths and constructed outfall drains for direct disposal to oceans or saline lakes, and evaporation ponds.

Another potential disposal option is deep-well injection. However, in California, this approach failed due to the slow permeability of the geologic strata from the plugging of conducting pores. Besides the natural drainage conditions, the suitability of each of these measures depends on: the quality and quantity of drainage water requiring disposal; environmental and health risks; available technology and resources; and economic considerations.

Drainage water disposal into natural surface water bodies should entail minimal deleterious impacts on other downstream water uses (agricultural, industrial and municipal) and wildlife. Disposal on land in closed basins and agricultural evaporation basins should avoid undue harm to the ecology, particularly aquatic biota including fish and waterbirds.

NON-PHYSICAL DRAINAGE WATER MANAGEMENT OPTIONS

In order to be fully effective, non-physical management measures, i.e. policy and legislation, should accompany physical drainage water management options. Different countries have implemented several policies, and these are often part of a general pollution policy.

One of the most frequently mentioned principles with respect to pollution is that of the polluter pays. This means that economic policy instruments are applied to change the behaviour of farmers in such a way that pollution is minimised or that at least the polluter pays for its effects. Weersink and Livernois provide an overview of such economic policy instruments for resolving water quality problems from agriculture. This overview was compiled for pollution by nutrients in humid temperate areas, such as Canada. However, some of these policies have relevance to pollution from the drainage of irrigated lands. The main problem of policy instruments in drainage water management is that water quality degradation from agriculture is non-point source pollution. Direct links between agricultural practices and water quality degradation are often difficult to determine and quantify.

Table 1. Economic policy instruments

Base	*Target*	*Form*
Performance of agricultural systems	Emission	Charges
	Ambient	Charges
		Salinity permits
Agricultural practices	Inputs	Charges
		Subsidies
	Outputs	Charges
		Subsidies

Weersink and Livernois divided the policy instruments into two groups: (i) those based on the performance of an agricultural system, i.e. on the actual amount of pollution that caused; and (ii) those based on agricultural practices such as fertiliser and agrochemical uses (Table 1). In the latter case, the underlying assumption is that certain agricultural practices will lead to more or less pollution. These policy instruments try to influence the practices and, indirectly, the pollution resulting from these practices.

Emission Levels

Fees are levied on the discharge of the polluter into the water. The aim is to stimulate the polluter to adopt practices that minimise pollution or to make the polluter pay for some of the damage caused. This requires measurement of individual discharges, which is often costly and not always practical to implement. Measurements at the individual farm level are probably only feasible with modern and large farms, such as in the United States of America and Spain. However, the case of the Panoche Fan shows that it remains difficult to establish equitable fines.

Discharges from upstream farms with natural internal drainage cannot be measured while they may contribute significantly to the pollution in downstream regions. An alternative to measuring and levying fees on individual farms is to implement it at district level or at that of a water users association. This has the disadvantage that farmers applying conservation measures may still pay for the pollution caused by badly managed farms in the same district.

When basing fines on emission levels, there should be a decision as to whether to base them on concentrations of polluting elements or on the total load (i.e. concentration multiplied by flow). Concentration is the water quality parameter used in drinking-water standards or for the health of aquatic animals and plants. Placing only concentration limi[illegible] on discharge might encourage dilution or inefficient water [illegible]se. Where both concentration and load limits are enforce[illegible] [illegible]y tend to promote efficient water use.

Ambient Levels

Basing charges on the pollution in the downstream receiving water body can sometimes be a feasible alternative to basing them on the emission level, i.e. where the source of pollution is extremely difficult to measure. For example, in the case of pollution of groundwater, individual contributions to the pollution are difficult to measure and thus difficult to quantify. A disadvantage of using ambient levels under these conditions is that all farmers, including those applying sound water management practices, are charged equally.

Another case where this policy measure may be an attractive option is where the carrying capacity of the receiving water body changes continuously. For example, in India during the monsoon, river discharge is high and effluent disposals during this period increase downstream concentrations only slightly. On the other hand, during the dry season when river flow is low, emissions affect water quality substantially. In this case, using the ambient criterion reflects the major concern: maintaining river water quality for downstream users. This method stimulates corrective measures during critical periods.

Salinity Permits

An interesting solution for salinity management on a regional level is that adopted in the Murray-Darling basin in Australia. States within the basin have to meet electrical conductivity (*EC*) levels at the end of their river valleys, this in order to maintain a favourable water quality in the entire downstream river. In order to reach this goal, a system of salt credits and debits is used. Credits are obtained for the implementation of any works that reduce the salinity in shared rivers.

Debits are incurred based on the estimated shortfall in protecting shared rivers. The balance of credits and debits is registered for each state, and as a general principle each state must be in credit. The credits and debits are converted to *EC* impact at a location in the downstream area of the basin. This method allows states and catchment management authorities to decide on the most cost-effective options for their area whilst contributing to the overall basin-wide river salinity management plan.

Charges on Inputs

Instead of charging for pollution outflow from irrigation systems, economic incentives can be directed at reducing the inflow, which will normally lead to a reduced pollution outflow. Thus, as a way of reducing pollution, input charges seek to conserve water and minimise the use of agrochemicals and fertilisers. Volumetric water pricing is the main way of providing an economic incentive to save water. However, research from Iran has highlighted some limitations of this approach.

In the Iranian case, to serve as a proper economic incentive to change to pressurised irrigation, the total costs of water would need to equate to 60 percent of the revenues of a wheat crop, while at the moment they stand at 5 percent. Such an increase is probably politically impossible and would reduce farm incomes drastically. Surface irrigation improvement is less costly than changing to pressurised irrigation. Therefore, volumetric water pricing may be an economic incentive to improve surface irrigation water management.

Charges on inputs should also consider the way in which farmers can reduce water consumption. The main changes farmers can make are: grow less-water-demanding crops; apply water saving technologies; and change over to rainfed farming. All these responses lead to less water consumption, but sometimes also to significant drops in farm income level.

On the basis of computer simulations, Varela-Ortega *et al.* analysed the responses of farmers in Spain to increasing water prices. Results showed that in modern irrigation districts, where irrigation efficiency is already quite high, no changes in water demand would occur unless the prices became restrictively high and farmers needed to change to rainfed farming. In contrast, in areas with the older irrigation systems, farmers would have ample scope for improving their water use efficiency.

An alternative to volumetric pricing is tiered water pricing. In this system, farmers pay a lower volumetric price for a reasonably necessary amount of water to grow a certain crop. A higher volumetric price is charged for additional amounts of water. Such a system is explicitly designed to reduce the amounts of water that generate most of the drainage water.

The reasonably necessary amount should include the minimal leaching requirement in order to maintain a favourable salt balance in the rootzone unless sufficient rainfall occurs in the area. A main problem would be to define the 'reasonable necessary amount of water'. In Broadview Water District, California, the United States of America, tiering levels differ only between crops and not between soils, because farmers consider the latter approach to be inequitable.

Tiering levels were established at 90 percent of the average depths applied in the years before the system was implemented. This system of tiered water pricing led to a decrease in water applied to five of the seven main crops. More importantly, the drained volume in 20 of the 25 subsurface systems in the water district decreased by 23 percent and the salt load declined by 25 percent.

Volumetric pricing policies are typically only applicable in on-demand systems. Other systems deliver fixed amounts of water to farms, and they fall outside the responsibility of the individual farmers. In such systems, more gains could be obtained by matching supply and demand more closely to each other in a technical/operational way.

Subsidies on Practices

A changeover to water saving or pollution prevention measures might not always be the objective of individual farmers nor might it always be to their advantage. In the case of Iran, it was calculated that the investments in sprinkler irrigation are of the same order of magnitude as the value of the water that can be saved, if this is used to irrigate a larger area or high-value crops. However, if the water saved is diverted to irrigate land in other areas, upstream farmers will be investing while benefits accrue to other areas. Another example is the case of farmers located in the higher areas with free drainage in the Panoche Fan. Farmers in these areas contribute much to the pollution, but there is no direct incentive for them to invest in water saving or pollution prevention measures. In both cases, subsidies might stimulate farmers to invest in water saving technologies or pollution prevention measures.

Charging/subsidising Outputs

A final instrument proposed by Weersink and Livernois is a charge on the crops. In the case of nitrogen leaching, the charge focuses on crops that need large amounts of nitrogen, and thus have a higher risk of causing pollution. In irrigated agriculture, one could think of policies that discourage growing crops with large water requirements (such as alfalfa) by charging the crops or by promoting deep-rooting crops or less-water-demanding crops through subsidies. A common example is where farmers pay a price per area depending on the crops grown.

Combined Measures

A single economic or policy instrument will probably not resolve water pollution problems originating from irrigated agriculture. Combinations of policies can help to enforce their effect. Especially in situations where farmers are inequitably charged, a combination of measures might help to charge farmers more equitably. For example, emissions from an irrigation district could be measured and fines applied where limits are surpassed. These fines could then be distributed amongst the farmers, applying a ranking based on how farmers have used their inputs or what measures they implement. The most inefficient users should then pay comparatively more than the most efficient ones.

SELECTION AND EVALUATION

Normally, more than one drainage water management option will need to be considered and implemented to attain the desired objectives. In this case, interactions and trade-offs occur. Because of the complexity of the processes and interactions, computer models are best suited to selecting an optimal combination of drainage measures.

WATER QUALITY CONCERNS

There are several factors to consider when determining the opportunities for and constraints on the safe use, treatment and disposal of agricultural drainage water. Information and data desired at the site of drainage water production include: rate of drainage water production per unit area, chemical concentration of constituents of concern, and the rates of mass emission.

Drainage water management requires additional information and data on drainage water quality and its suitability for the intended water uses as well as an understanding of environmental and health concerns. Upstream drainage water management affects the needs and water quality requirements of downstream water users.

Drainage Water Quality

Table 2 provides a summary of changes in water quality expected in irrigation return flow relative to irrigation water applied. The expected differences in quality in the return flow are described relative to the supply water because actual concentrations in supply waters vary. The operational spill waters (bypass water) from distribution conveyances are not expected to differ much from the quality of the supply water except for some pickup or deposition of sediments.

Table 2. Expected quality characteristics of irrigation return flow as related to applied irrigation waters

Quality parameters	*Operational spills*	*Irrigation tailwater*	*Subsurface drainage*
General quality	0	+	++
Salinity	0	0, +	++
Nitrogen	0	0, +, ++	++, +
Oxygen demanding organics	0	+, 0	0, -, -
Sediments	+, -	++	-
Pesticide residues	0	++	0, -, +
Phosphorus	0, +	++	0, -, +

In contrast, surface runoff or irrigation tailwater tends to pick up considerable amounts of sediments and associated nutrients, phosphorus in particular, as well as water-applied agricultural chemicals such as pesticides and nitrogen fertilisers (especially anhydrous ammonia). Tailwater is typically similar to the applied waters in salinity and oxygen demanding organics, termed biochemical oxygen demand (BOD).

Subsurface drainage is enriched in soluble components such as dissolved mineral salts and nitrates, very low in sediments, whereas other quality parameters are similar to the irrigation water. These changes in water quality of irrigation return flow depend on a number of factors including irrigation application methods, soil properties and conditions, application of agricultural chemicals, hydrogeology, drainage system, climate, and farmers' water management.

In many regions of the world, municipalities and industries discharge wastewater into open drains initially intended for the conveyance of only agricultural drainage and storm water. In developing countries especially, municipal and industrial wastewater is often insufficiently treated before disposal into such open drains. The result is a risk that agricultural drainage water quality might be seriously contaminated with microbes, pathogens, toxic organics and trace elements including heavy metals.

A knowledge of the composition of the drainage effluent and the ability to predict changes in the composition as a result of changes in crop, irrigation or drainage water management practices are important in the planning and management of drainage water.

FACTORS AFFECTING DRAINAGE WATER QUALITY

Geology and Hydrology

The geology of the region plays an important role in drainage water quality. Through weathering processes, the types of rocks (both primary and sedimentary) in the upper and lower strata define the types and quantities of soluble constituents found in the irrigated area. The oceans have submerged many parts of the continents during a period in their geological history.

The uplift of these submerged geological formations and receding seas have left marine evaporites and sedimentary rocks behind, high in sea salts including sodium, chloride, magnesium, sulphate and boron. These geological formations exist in varying thicknesses, depths and extents on the continents. Through hydrological processes, solutes can enter the upper stratum by irrigation or floodwater, upward groundwater flow in seepage zones, with rising groundwater levels, or capillary rise.

Once the solutes are in the upper strata, they influence the quality of agricultural drainage water through farmers' irrigation and drainage water management. The following example shows how the geology and hydrology of an area influence the quality of agricultural drainage water. It also illustrates the relationship between geomorphology, waterlogging and salinisation.

The eastern groundwaters are characterised as low-salt calcium-bicarbonate-type water with total dissolved solids (TDS) typically in the 200-500 mg/litre range. In contrast, the soils on the western side were formed from alluvium of the Coast Range made up of uplifted marine sedimentary rocks. The soils on the western side tend to be finer textured and saline. The groundwaters on the western side are characterised as moderately saline sodium-sulphate-type waters with TDS typically in the 1 000-10 000 mg/litre range.

The unconfined aquifer in both sides of the valley is gradually being filled up with decades of irrigation deep percolation. The soils in the valley and lowest part of the alluvial fans in the western side are waterlogged and salt affected. A nearly water-impermeable clay layer known as the Corcoran clay, about 200 m deep, serves as the boundary between the unconfined and confined aquifer. The groundwaters in the confined aquifer contain from 500 to 1 000 mg/litre TDS. During the geologic past, plate tectonics caused the horizontal-lying Corcoran clay in the shallow sea to tilt upwards forming the Coast Range.

The applied irrigation water is about 450 mg/litre calcium-bicarbonate-type imported water from the Sacramento River basin to the north. Much of the surface runoff is captured and reused on site. Much of the collected saline subsurface drainage water (4,000-10,000 mg/litre *TDS*, sodium-sulphate type) is discharged into the San Joaquin River. Especially high concentrations of trace elements such as boron and selenium originating from the marine sedimentary rocks, found in the subsurface drainage waters, have given rise to environmental and health concerns. Discharges from these areas are now constrained by waste discharge requirements.

A second example comes from the Aral Sea Basin, where in total 137 million tonnes of salt are annually discharged, of which 81 million tonnes (59 percent) originate from the irrigation water and 56 million tonnes (41 percent) from the mobilisation of salts from the subsoil. In the mid-stream areas, mobilisation of subsoil salts is the most substantial. The annual discharge from the Karshi oblast (Uzbekistan) is 10.8 million tonnes of salts, which corresponds to about 34 tonnes per hectare. About 4.3 million tonnes (about 40 percent) originate from irrigation water and the remainder are mobilised salts from the subsoil through irrigation and drainage. Further, in Australia, large amounts of salts are added to the soil profile by atmospheric deposition of salts from upwind wind erosion of salt pans.

SOILS

Soils serve not only as a medium for plant growth but also store water and nutrients and serve as the porous transport media. The soil's eroding capacity and chemical weathering leads to the generation of water-borne suspended particles and solutes, ranging from nutrients to all kinds of contaminants. Therefore, to understand the role of soils on drainage water quality, it is necessary to understand water movement over and through the soil and the associated suspended and dissolved substances it carries.

Of the water added to the soil, either in the form of rainfall or irrigation, part is lost through runoff and direct evaporation at the soil surface. Runoff water collects in natural and constructed surface drains from where it finds its way to the final disposal site (a river, evaporation pond or outfall drain to the ocean or saline lake). The other part infiltrates into the soil. This water fills up the soil pores and restores the soil moisture content up to field capacity under free drainage.

The stored water is now available for plant root extraction to satisfy the water requirement of the crop. Any water in excess of field capacity percolates below the rootzone to greater depth in the vadose zone. The deep percolation water may eventually serve as recharge to the groundwater or saturated zone. In irrigated areas with shallow groundwater tables, the recharge is immediate and causes the water table to rise. Where subsurface drainage is installed in waterlogged soils, the drainage system removes deep percolation and groundwater. Where the soil moisture content in the rootzone drops as a result of evapotranspiration and if there is no recharge from irrigation or rainfall, capillary rise into the rootzone might occur, depending on the water table depth, soil texture and structure, and seepage.

Runoff in irrigated agriculture is mainly related to the intensity of irrigation and rainfall events in comparison to the infiltration capacity of the soil. Where the infiltration rate is smaller than the irrigation or rainfall intensity, water will accumulate on the soil surface and run off under a minimum surface slope.

Soil degradation, in terms of compaction and crust formation as well as cultivation on steep slopes, promotes surface runoff. Through the physical forces of the running water, soil particles become suspended in the water and are transported to open drains, ditches, streams, rivers and lakes. Deposition of suspended sediments may occur downstream when current velocities decrease. Suspended soil particles are harmful to aquatic life as they diminish light transmission, but also because chemical contaminants may be associated with suspended sediments. Degradation of drainage water quality as affected by runoff from agricultural land is especially important in hilly areas and in areas with excess rainfall.

In the more arid areas and in flat plains, water flowing through the soil profile and associated solute fluxes affected by mineral solubility and adsorption processes are of more importance for the final quality of the drainage effluent than surface runoff. In some places, weathering of soil particles might play a major role in the quality of drainage water (such as dissolution of gypsum). However, in general the soil's ability to adsorb and release through ion exchange and transform chemical elements through microbially-mediated redox reactions plays a more important role.

In this context, soil characteristics such as water holding capacity, hydraulic conductivity, clay and organic matter content, soil minerals and soil microbes are important characteristics. Soil degradation in the form of erosion, compaction and loss of biological activity reduces the water and solute holding capacity of the soils. This increases the mobility of solutes through the soil and increases the risk that pollutants such as salts, nutrients and pesticides will be lost both to groundwater and through interception by subsurface drainage to surface water.

CLIMATE

As the major transport of solutes through the soil is by the movement of water, climate plays a major role in determining drainage water quality. In humid tropics and temperate regions, the dominant movement of water through the soil is vertically downwards. Solutes, which are brought onto the soil by farmers or are naturally present in the upper soil layers, are leached into deeper soil layers and groundwater.

Conversely, in arid climates where evaporation largely exceeds precipitation the dominant water movement through the soil is vertically upwards except during rainfall or irrigation events. Therefore, the chemical composition of deeper soil layers influences the quality of the shallow groundwater and the composition of the soil moisture in the rootzone. Climate and temperature also play a role in the rate of weathering and chemical processes.

Cropping Patterns

Cropping patterns play an important role in the quality of

drainage water in a number of respects. First, crops extract water from the rootzone resulting in an evapoconcentration of salts and other solutes in the soil solution. Where the solubility product of minerals is exceeded through evapoconcentration, minerals precipitate out. This changes the composition of the soil solution and thus influences the chemical quality of subsurface drainage waters.

Second, crop residues add organic matter to the soil profile. Organic matter in the soil increases the adsorptive capacity for metals and other solutes. Furthermore, organic matter enhances the soil structure, which increases the water holding capacity of the soil. The organic matter also serves as a carbon source for soil microbes involved in transformations such as denitrification, sulphate reduction and methane production in submerged soils. Third, plants extract nutrients through their rooting system and some plants have the capacity to accumulate large amounts of certain salts and toxic elements.

Fourth, as not all crops have the same salt tolerance the type of crop largely determines the maximum salt concentration in the rootzone and the amount of water needed to maintain a favourable salt balance in the rootzone. Last, incorporation of nitrogen fixing crops such as legumes can help to reduce nitrogen leaching. Legumes in symbiosis with nitrogen fixing bacteria are both users and producers of nitrogen. They can substitute chemical nitrogen fertiliser in the crop rotation. Deep-rooted perennial crops such as alfalfa can also help to prevent nitrogen leaching by absorbing large amounts of nitrogen.

Use of Agricultural Inputs

Application of fertilisers, pesticides, soil and water amendments, and animal manures may influence the quality of drainage water to a great extent. The amounts and timing of application in relation to the growing stage of the crops, timing of irrigation, drainage practices and applied soil conservation measures largely define the influence of fertiliser, amendment and pesticide application on drainage water quality. Furthermore, the characteristics of the fertilisers themselves play a major role, also on the possible contamination of drainage water.

Most nitrogen fertilisers are highly soluble and mobile in the soil, and nitrates readily enter drainage water through leaching processes. A portion of the nitrogen fertilisers made up of ammonium or anhydrous ammonia is initially adsorbed to the soil exchange complex but ammonium ions oxidise readily to nitrate. This is also true of urea containing nitrogen fertilisers that eventually oxidise to nitrates. Conversely, phosphorus fertilisers are less mobile in the soil because they have very low solubility, and phosphates are adsorbed on positive sites in soil organic matter and clay minerals.

The main route for phosphorous to drainage water is through runoff as sediment-bound inorganic and organic phosphate. Runoff waters may contain residues of anhydrous ammonia injected into irrigation water. Ammonia is highly toxic to fish. Runoff waters may also contain sediment-bound organic nitrogen. Excessive levels of nitrogen and phosphate in discharge waters may result in the eutrophication of water bodies.

Water and soil amendments such as gypsum may contribute significantly to salinity. Although gypsum is sparingly soluble, calcium contributed from gypsum may exchange with adsorbed sodium, and sodium sulphate is a highly soluble mineral. Acidic amendments react with soil calcium carbonates but do not contribute to salinity because the solubility of calcium carbonate is very low. Where animal manures are mineralised, nitrates and salts are produced. Some animal manure contains appreciable amounts of salts.

Irrigation and Drainage Management

Irrigation and drainage management are the main factors influencing the flow of water over and through the soil in arid zone croplands. As solute transport takes place mainly through soil water fluxes, irrigation and drainage management determine to a great extent the solute fluxes through the soil. In irrigation management, timing in relation to crop water requirements and to fertiliser and pesticide application is key to controlling the amount of soluble elements that will leach below the rootzone, from where they can be intercepted by subsurface drains. The timing of irrigation also affects capillary rise into the rootzone, which might cause an accumulation of salts in the rootzone but at the same time reduces the need for irrigation water. Of equal importance to the timing is the amount of irrigation water applied. Excess water, including infiltrated rainfall, leaches to deeper soil layers.

Drainage Techniques and Design

The choice of drainage technology and certain design choices influence the quality drainage effluent considerably. Subsurface drainage enhances the flow of water through the soil. Studies from the United States of America have shown that, in comparison to soils drained by means of surface drains, subsurface drainage reduces the amount of runoff and subsequently reduces the phosphorous contamination of surface water.

At the same time, subsurface drainage enhances nitrogen leaching. Upon interception of leaching water, the nitrogen concentration of the final drainage effluent increases. In the same way, subsurface drainage intercepts other soluble elements present in the soil. This offers the possibility to control the concentration of harmful salts and toxic trace elements in the rootzone for optimal crop production but reduces the possibility for disposal and reuse of drainage water.

For water table and salinity control in irrigated lands, two types of drainage technologies are generally employed: subsurface tile drainage, and tubewell drainage. Subsurface tile drains are generally installed at a depth of less than 2 m below the soil surface. However, exceptionally, drains are installed at depths of up to 3 m. Tubewell drains vary between 6 and 10 m but might reach a depth of 100 m, e.g. deep tubewell drains in Pakistan. These differences in drain depth influence the quality of the effluent.

The quality of drainage water from subsurface tile drains is influenced by the quality of irrigation water, the applied farm inputs and the quality of the shallow groundwater. In contrast, the quality of drainage effluent from tubewell drains is related mainly to the quality of the groundwater and to a lesser extent to the quality of irrigation water. Based on computer simulations in Pakistan, a comparison of tubewell and tile drainage showed that for Kairpur the salinity of drainage effluent was 11 750 and 2 100 ppm for tubewell and tile drainage, respectively, whereas for Panjnad Abbiana it was 22 918 and 610 ppm, respectively.

Table 3. Salt applied in irrigation water and removed by drains

Treatment	Salt removed (kg/ha)	Salt removed/Salt applied
Deep drains	607	11
Shallow drains	32	0.7

The depth of tile drain installation also influences the amount and quality of the generated drainage water. In the Aral Sea Basin, deep drains increase the flow that has to be evacuated and its salt content through mobilisation of fossil groundwater through deep flow paths. For example, drainage water in the Karshi oblast collector is extremely saline (up to 8 g/litre). Expected mineralisation levels of drainage water resulting from irrigation only are much less.

Experiments in Australia with shallow and deep drains in a vineyard on a clay soil led to the same conclusion.

Less drainage effluent was generated with shallow drains (0.7 m deep) than with deep drains (1.8 m deep). Moreover, the shallow drainage water had a lower salinity than the deep drainage water; together with reduced drainage volumes this resulted in a reduced salt load. As the deep drains removed much more salt than imported with the irrigation water (Table 3), this led to the conclusion that the salinity in the effluent of the deep drains was derived from deeper soil layers.

Model simulations by Fio and Deverel confirm these observations. They showed that the base flow towards tile drains increased and the quality decreased with increasing installation depth under non-irrigated conditions. They also showed that this is a result of the path flows becoming deeper and longer with increased drain depth and spacing. On the other hand, during irrigation events the discharge of the shallow and deeper drains became similar as recharge due to deep percolation increased and the proportional contribution of deep groundwater to drain lateral flow decreased.

REFERENCES

Breininger, D. R., Provancha, M. J., and Smith, R.B., "Mapping Florida scrub jay habitat for purposes of land use management," *Photogrammetric Engineering and Remote Sensing*, 57, 1467-1474, 1991.

Galli, J., *Peat-Sand Filters: A Proposed Stormwater Management Practice for Urbanized areas*, Metropolitan Washington Council of Governments, 1990.

Julien, P. Y., "Transforms for Runoff and Sediment Transport," *J. Hydr. Engr.*, ASCE, 1(3), 114-122, 1996.

Leitch, C., "In Review, Impacts of Land Use Change on Freshwater Runoff into the Near-Coastal Zone, Holetown Watershed, Barbados: Comparisons of Long-Term to Single-Storm Effects," *Journal of Soil and Water Conservation*, 1996.

Minner, M., Harbor J., Happold S., and P.Michael-Butler, "Cost Apportionment for a Storm-Water Management System," *Applied Geographic Studies*, 2 (4), 247-260, 1998.

Stahre, P. and Urbonas, B., *Storm-Water Detention for Drainage, Water Quality and CSO Management*, Prentice Hall, Engelwood Cliffs, New Jersey, 1990.

Woolpert Consulting, *Cuppy-McClure Watershed Storm Water Management Plan and Celery Bog Park*, Woolpert Consulting Company, Indianapolis, IN, 1995.

10

Capacity Building in Irrigation and Drainage

Capacity building is not something new; it has been a leading issue in development for many years and comes from the widely acknowledged shortcomings in development assistance over the past 50 years. However, what is new is the attention that capacity building is now receiving in the irrigation and drainage world where it is becoming an issue in its own right rather than being embedded in infrastructure investment projects.

The need for capacity development in irrigation and drainage became particularly obvious where irrigation management transfer programmes were implemented and when people realised that transferring responsibilities without transferring/raising the capacity was a major threat to the success of this institutional reform.

Many developing countries now say they need more capacity, and aid agencies are keen to supply it or at least help to create it. However, it remains a concept of enormous generality and vagueness wrapped up in a host of concepts such as participation, empowerment, technical assistance and organisational development. Moore states that: "Capacity building includes everything that was covered by the different definitions of institution building and more besides...Aid agencies would be wise to have no truck with the new jargon of 'capacity building' and to insist on using language and terms that have identifiable and precise meanings." Others refer to the weary mantra of capacity building that is not leading anywhere. The calls for capacity building in irrigated agriculture suffer from these same vague generalities.

Many senior government officials talk about a lack of capacity constraining development and see solutions in terms of more infrastructure and training for farmers and local professionals rather than in reforming the broader institutional context in which they function. Unfortunately, donors have tended to reinforce this approach. They have shaped technical cooperation by concentrating on new works and providing technical assistance to fill the skills gap and to pass on know-how rather than focus on national priorities. Disappointingly, such approaches have failed to yield the expected benefits. So beyond the general idea that more capacity is required, there is a need for greater clarity of thinking about the practicalities of planning, designing and implementing capacity building.

CAPACITY BUILDING

Capacity and Development

The established model for capacity building developed with the advent of the aid era following the Second World War. The Marshall Plan, introduced to support the rebuilding of Europe, was very successful but it generated an overly simplistic and optimistic view of what worked; transfer capital and know-how to other countries and swift economic development would follow. Subsequent experience showed that this view underestimated the importance of local knowledge and institutions in the process of economic development and was compounded by aid driven by politics rather than results. Aid was then criticised for undermining local capacity rather than building it, ignoring local wishes and favouring high profile activities.

In the past decade, there has been a change in emphasis towards seeing development as "locally owned" so that cooperation does not seek to do things for developing countries but with them. In support of this Fukuda-Parr, Lopes and Malik point out that most countries evolved organically, building their own capacities, and so the assumption that developing countries should simply start from someone else's blueprint files in the face of history. They suggest that countries should build on the wealth of local knowledge and expand these to achieve whatever goals and aspirations the country sets itself.

They use Morgan's description of capacity simply as the ability to perform functions, solve problems, and set and achieve objectives, and raise the issue of national capacity being more than the sum of the capacity of individuals, which points to the importance of people working in organisations and networking. This has become known as social capital and in its crudest terms it means the more people trust each other, the better off their society. Economists now recognise this as an addition to natural resources, the rule of law and the market that they believe guide the hand of selfish human actions to serve the common good. Traditional rice-based systems required high social capital for the massive labour needed to build and maintain the terraces and to synchronise cropping patterns for effective water and pest management

This change in development has also had a profound influence on irrigated agriculture. In addition, there is a growing understanding that successful irrigation farming requires much more than investment in construction and training. Farmers are being encouraged to take on more responsibility for water management, and they are increasingly being exposed to private sector market forces. To cope with these and other pressures, farmers are beginning to demand better support services, advice, access to new technologies, finance, equitable and fair water management regulations, and legal redress when things go wrong. These are seen collectively as a lack of "capacity" and the need is to build or develop it so that the water sector can function properly and poor farmers can take on more responsibility for their own development.

Browne suggests that capacity building is both easy and hard to define. The easy part is the generic definition about skills and capabilities - the ability to perform functions, solve problems and set and achieve objectives - and is a significant step from earlier definitions that looked only at human resource development. However, the difficult part is answering the questions about which skills and whose capability, and this raises many more issues about the complex nature of development.

Alaerts examined at several definitions applied to the water sector but avoided recommending one, as all have limitations and focus attention on different aspects of capacity building. Those that attempt to be all-inclusive tend to become rather turgid and incomprehensible. So, is there an acceptable definition for the irrigation community?

The definition formulated by the United Nations Development Programme comes close. It is general, wide ranging and encompasses the approaches to development that are now well accepted in irrigation. It states that: "Capacity building is the sum of efforts needed to nurture, enhance and utilise the skills and capabilities of people and institutions at all levels - locally, nationally, regionally and internationally so that they can better progress towards sustainable development". It continues by saying: "at the basic conceptual level, building capacity involves empowering people and organisations to solve their problems, rather than attempting to solve problems directly. When capacity building is successful, the result is more effective people and institutions better able to provide products and services on a sustainable basis."

End Products

The "end products" of capacity building are trained individuals, established irrigation organisations, new water laws, etc. However, it is increasingly recognised that great care must be taken over their production if they are to be useful. The top-down approach of government and aid agencies deciding what needs to be done and how it should be done is now discredited. There is a realisation that products need to be produced by working in response to local initiatives with local ownership and leadership and incorporating the ideas of participation and subsidiarity. Thus, the process of building capacity becomes an essential part of producing a useful end product.

Eade makes this point by saying that capacity building is an approach to development, not something separate from it and not a discrete or prepackaged technical intervention intended to bring about a predefined outcome. It is the reason why many people now use the term capacity development rather than capacity building. It transfers the emphasis from the end product to the process of achieving it.

External/Internal Balance

It is difficult, if not impossible, to separate the internal nature of capacity development from external aid and technical cooperation. Any discussion about capacity development inevitably leads to how it will be financed and what specialist help is needed that is not available in-country. With its emphasis on external support, the Marshall Plan set the early agenda, but this agenda is now changing as both the approaches to aid and technical cooperation are being reshaped to fit more closely with the new processes of capacity development.

The current external/internal balance is weighted heavily towards internal actions but there is an acceptance that outside help can support and accelerate the process. An outsider can often help to focus attention on the big issues, and the stimulation that external technical cooperation encourages can help to speed up processes and broaden thinking. They can provide a useful role in helping to identify capacity constraints and support both government and the private sector to take action. The key is to establish just how much assistance is needed and can be provided to support the internal process without upsetting it and jeopardising its sustainability.

Long-term partnerships are now seen as the essence of good collaboration between donors and recipients and not the more traditional North-South dialogue. Such relationships can influence the process positively as trust and effective work relationships develop over time. However, striking the right balance is the greatest challenge when one party is seen to be giving and one is receiving. However, some still view any external support as evidence of a lack of sustainability.

Framework

The UNDP definition is one step in the process of clarifying what capacity building means. However, it does little to describe the territory of capacity building in irrigation, how to assess the capacity gaps and what practical steps are needed to overcome them. Help with describing the territory comes from social science and the extensive conceptual thinking about capacity building that has been taking place over the past decade on development generally. This has produced a generic framework that has become widely accepted in development and is also gaining acceptance in the water sector. It is suggested that this is a useful framework for irrigated agriculture.

Bolger describes the generic framework in terms of different levels of capacity, namely: individuals, organisations and societies, but with an emphasis on the links between them. For example, the performance of a WUA is shaped as much by society (laws and regulations) as it is by individuals (skills, leadership and relationships).

Four levels of capacity emerge from this approach. Each represents a level that can be examined and analysed as well as a possible entry point for support from a donor or technical cooperation. Interpreting this framework in terms of irrigated agriculture, the four levels are:

— *Individual level.* This is the most "concrete" and familiar part of capacity development and includes the education and training of the various stakeholders, farmers, local professionals, engineers and other disciplines involved in irrigation.

— *Organisational level.* This refers to groups of people such as water user organisations, research groups, government extension agencies, private companies that share common objectives such as improved livelihoods at a farming level, improved water management or increased agricultural productivity at a national level. Institutions are the rules and agreements, formal and informal, and shared values that bind organisations. Thus, the capacity of an organisation is embedded in the ability of its individuals to work together within established rules and values.

— *Sector/network level.* The sector level emphasises the point that irrigation is part of the larger picture of integrated water resource management and reflects the increasing awareness of the need for policies that integrate and cover all aspects of the water sector and not just irrigation, water supplies and the environment in isolation. The addition of networking at this level is less important for irrigation as this can take place at all the levels.

— *Enabling environment.* This represents the broad national and international context within which irrigated agriculture can develop, and this has considerable influence over what happens at the lower levels. It is concerned with policy at the highest levels in government and the ability of people at the lower levels to influence it. It is also about the socio-economic conditions that enable or discourage irrigation development, and the legal framework that provides farmers with security of tenure for land and water and the power to seek legal redress when contracts are broken.

Framework for Irrigated Agriculture

The generic framework provides a useful starting-point for mapping the territory of capacity development in irrigated agriculture. However, it needs to be enriched with specific aspects of social science, the practical activities of a well-functioning irrigation sector, and the activities that develop capacity. There are many irrigation-related issues to consider. The question is what to include and what to leave out. Adding too many can lead to confusion and paralysis, whereas adding too few can lead to an oversimplification that does not bring out the main issues.

A compromise was proposed by introducing a second dimension based on the familiar activities of planning, design, construction, and O&M. In addition to these, research, education and training, and networking were added as important areas of capacity development in their own right. Research here refers to the capacity to develop new knowledge in a specialist institute or university, and education and training refers to the capacity to disseminate knowledge in colleges, universities, etc. Networking refers to the various formal and informal networks that are instrumental in disseminating knowledge and innovation that can and should take place at all levels of capacity development.

The historical development of agriculture in many countries has shown that the capacity of farmers' unions and cooperatives to communicate with one another and with their members has been an important tool for boosting agricultural progress. Similarly, engineers and agriculturalists have benefited from the networking capabilities of their professional organisations. Guiding principles of subsidiarity and participation were added to demonstrate the move away from the traditional top-down approach to development.

Each box in the framework represents an activity in the water sector that requires trained individuals working in organisations and within an enabling environment. If constraints are identified for a particular activity, then the need can be assessed, actions taken and impacts measured. However, as Bolger points out in relation to the generic framework, the levels (or boxes in this case) are not simple watertight compartments but are very much linked together is a variety of ways that depend on local circumstances. The simplicity of the framework should also not mask the complexity of the issues to be dealt with.

Developing a Strategy

How useful is such a framework? In the same way that Bolger's framework provides a generic "picture" of capacity development, it can be argued that this modified framework provides a similar "picture" for irrigated agriculture. It covers the territory in which capacity is needed; it can provide a convenient starting-point for questions and discussion at both local and national level, and be helpful in talks with donors in establishing clarity of purpose on areas of constraint and what needs to be done: Where are the capacity gaps? Is it individuals and if so which? Are the constraints mainly in O&M on farms or in support organisations provided by government or the private sector? Does the socio-economic environment discourage irrigation or is the education and training provision unable to provide the basic human resources on which irrigation can grow?

Thus, the framework can provide a starting-point for discussions that can begin to provide answers to these questions. The priority given to each box will vary considerably from one place to another depending on local issues such as the way irrigation is practised (e.g. privately-run small holder irrigation, publicly-owned large irrigation and drainage networks and commercial farming) and the need to solve local problems (e.g. salinity and water logging, inadequate education facilities, and irrigation management transfer).

Strategic Planning

This is the beginning of a strategy development process for capacity development that includes:

- diagnosis: identification of the gaps;
- definition of the goals;
- means;
- monitoring of performance.

The CGIAR Secretariat defines strategic planning as: "a process by which an organisation develops the most desirable vision of its future, taking into account the constraints the organisation is likely to work within, and shows how the organisation can realise that vision." To date most capacity development "strategies" have tended to be a "bolt-on" to infrastructure development. They are often limited in scope, poorly defined, and developed with minimal stakeholder consultation. Thus, applying strategic planning to capacity development as the primary issue has little precedence.

However, this approach has served well in other sectors and so it is suggested that it is a viable approach for irrigation. It is now being advocated strongly by the

IPTRID, whose main thrust is to support strategic planning, design and implementation of capacity development programmes and projects in developing countries. It is unlikely that such a strategy can be divorced from infrastructure but it is a question of emphasis. If capacity development is the key issue, the question then becomes: what infrastructure is needed to support it?

The IPTRID proposes an initial needs assessment at regional, national and subnational levels in order to examine the present situation and to answer the question: where do we stand? The aim is to identify the priority technical issues in terms of water conservation, water productivity, land degradation, salinity, etc. and the existing institutional capacity and weaknesses at strategic and operational levels.

Roundtable conferences and workshops are proposed to ensure full participation of all the stakeholders and ensure that their views are taken into account for the next step, which is to answer the question: where do we want to go? A strategy can then be developed to reach the desired goals - how do we get there?

Projects to implement the strategy are the next logical step, but there is a divergence of thinking. Is it possible to have capacity building projects designed to support what is essentially a process? Projects are usually end-product oriented. They have specific targets to be met in a given time frame with strict budgets and control on what and how money is spent. This would seem to be at odds with current thinking on capacity development, repeating here the point made by Eade that it is "not a discrete or prepackaged technical intervention intended to bring about a predefined outcome." Thus, can a project, with all its tools for project management and meeting deadlines and objectives, be applied to capacity development? If projects continue to be the way of supporting development, then some imaginative thinking will be needed in order to ensure a good marriage between them. This also has implications for aid funding, which is usually tied to projects. However, some donors are experimenting with alternative and more flexible aid packages that can take account of the flexible nature or processes as well as focusing on the end product.

An Agenda for Discussion

The framework provides a starting-point for planning a strategy for capacity development, but what are the basic issues that form the agenda for stakeholder discussion? This will vary from place to place for a wide variety of reasons, but some of the more common and important issues that are prevalent in many countries are raised here. The framework is used to provide a structure for highlighting these issues.

Individuals

There are many ways in which individuals can raise their capacity through education, training and community development, and participatory self-capacity development. It may be short-term, on-the-job training for farmers who need water management skills or to understand the workings of a WUA or more substantial long-term training for professionals, who undertake research, design irrigation works or provide advisory services.

Unfortunately, this is usually the only level at which capacity development is interpreted, often in complete isolation from the other levels. This should not cause too much surprise because training lends itself to specific and identifiable "concrete" actions that are more difficult at the higher levels. Some trainers would argue that everything eventually boils down to a training need even where it is an organisational deficiency. Training not only helps to improve practical skills and knowledge, it can also help change attitudes in the workplace.

Farmers usually enter the irrigation world through practice on the farm, whereas professionals usually come via specialised courses in colleges and universities where individuals have chosen a career in agriculture or engineering.

Training

Short-term training is usually geared to the acquisition of more immediate skills for well-defined tasks, such as managing canals, designing pumping stations and organising WUAs. It may be on-the-job training or special courses on particular issues for a wide range of people from farmers to professionals. They may be conducted informally on-site or more formally at a training centre or college, in-country or abroad.

Increasingly short-term training is also seen as a means of changing attitudes as both farmers and engineers take on new roles in the process of irrigation management transfer, and there is a change in the objective of training from mere knowledge transfer towards increasing problem solving capacities. Competences are a complex interaction of knowledge, skills and attitudes, and some programmes are now using participatory training methods to engender a continuing process of participation in subsequent water management practices. This is particularly important where individuals form groups or organisations such as WUAs to undertake joint water management functions.

Short-term training in irrigation and drainage is not without its critics. There are many examples of training courses that are not effective in matching demand with supply in a timely manner. Typical issues raised include: inappropriate skills and knowledge being provided; poor training methods; overly theoretical curricula lacking practical application; and poor course timing so that skills learned are not used or needed. It is not uncommon for attendance at a training course to be seen as a reward for good service, or it is this or that person's turn to go rather than an opportunity to acquire new and useful skills. Biswas provides a number of examples of irrelevant, inappropriate and expensive training that was either too general or not related to the job, and suggests that this should be a major cause for concern.

Typically, where donor aid is involved, short-term training is part of the investment package in order to provide the skills specifically for operating and maintaining a scheme. Although such training can be very effective for the project, it is often criticised for being narrowly focused and not giving adequate attention to organisational issues and the broader context in which irrigated agriculture functions.

Investing in people

Although few would doubt the value of educating and training individuals, the paradox is that few are willing to invest in people as enthusiastically as they invest in concrete. In spite of the rhetoric, the development of individuals is not receiving the type of support it needs and deserves. The mobility and fragility of labour are perhaps the main reasons for this. Irrigation schemes do not move around but people can and do. They also become ill and some die.

There are also commonly-held views about professionals who received large sums of money to train abroad in irrigation only to return home and go into construction work or banking. Although there may be an element of truth in this, it is unfortunate that this often dominates the level of perception and debate when the education and training of professionals is being discussed. It is seen as a waste and a reason for not continuing to train others rather than as a contribution to national capacity. Critics seem to give little thought to the very systems of education and training from which they themselves have personally gained great benefit in career development, personal choice and mobility of employment.

Human resources planning

Human resources planning is worth mentioning in the context of training individuals, although it cannot be divorced from the other levels of capacity development. It refers to a methodology developed in the 1970s and 1980s to determine future training needs in the labour market. It enabled planners to determine the numbers of people and the skills needed to bridge the gap between labour supply and demand in an industry or sector. This is not an exact science because of the problems of forecasting the future and, therefore, what labour demands are likely to be.

All studies made in the 1970s and 1980s recognised the importance of predicting the numbers required (quantity), ensuring they were well trained (quality), and that people worked within a sound institutional environment (organisations). However, there was a tendency in many countries to remedy deficiencies in one aspect by substituting improvements in another. For example, overstaffing was common with under qualified people working within poorly-managed organisations.

Substituting more untrained people to try and make up for a lack of quality is unlikely to work. Today the situation is completely reversed as a consequence of extensive irrigation management transfer. There are many WUAs or federations of WUAs that are critically lacking skilled staff in many aspects. There are also many irrigation agencies that are critically lacking skilled engineers able to understand and master, in their own context, the principles of modern water control and services.

Although human resources planning is not so popular at the moment, it does have the advantage of focusing attention on the issues of numbers, skills and education and training provision. It is unfortunate that the current literature on capacity development tends to ignore these basic issues. In the current surge of interest in the more complex institutional issues, there is a danger that these basics are being neglected. There appears to be a tacit assumption that the main skills of irrigation are being well provided when this is not the case.

Organisations

A wide range of organisations are important to the success of irrigated agriculture. At farm level, there are water user groups that share and manage common water interests technically, financially and legally. Other organisations, typically government run, are mandated to provide support services for farmers. In some countries, the private sector is also increasingly making an impact, for example through smallholders using low-cost affordable technologies and consultancy.

Support services

Farmers need organisations to provide advice and support. Traditionally, government has provided these in the form of extension services that link research with practical farming. However, there are some places were the private sector is beginning to play a role. This is particularly the case in private farming, where large farms require specialist services and smallholders need to acquire new skills to keep up with the dictates of the market.

On the whole, supporting organisations do not serve irrigators well. Government agencies are still the principal source of knowledge and skills. However, they tend to concentrate on crops, fertilisers and pesticides rather than on water, even in irrigated areas. They tend to be underfunded and staffed by inexperienced people who do not have the logistical support to reach the farmers. Where there have been successes, they have usually been underpinned by external aid, and thus there are question marks about their sustainability once the support ends.

Little has been documented and published about the role and function of irrigation advisory services and more specifically about how well they perform. There are few guidelines for others to follow. This situation has not been helped in recent years by the trend of transferring irrigation management responsibilities from government to farmers and the increasing amount of irrigation advice being organised outside the traditional government agencies by NGOs that do not normally have the resources to publicise their experiences. This lack of information can mean that it is difficult to determine what services are available in a country and how well they are working, and adds to the difficulty of planning a strategy for capacity development.

Organisational linkages

Government ministries involved in irrigation such as agriculture, environment, and water resources are often criticised for a lack of cooperation in what is essentially an interdisciplinary activity. Although they usually have common objectives, they tend to work independently. A typical division of responsibility is between the planning, design and operation of irrigation systems, usually the responsibility of a Ministry of Irrigation or Water Resources, and irrigated agriculture, usually the responsibility of a Ministry of Agriculture.

The former is staffed by engineers, who have little knowledge of crops and farming, and the latter by agriculturalists, who have little knowledge of engineering and hydraulics. The sad fact is that most of those involved do not see the need for such knowledge nor for closer linkages with each other. As one senior irrigation consulting engineer put it: "My engineers do not know the difference between wheat and rice - but the really sad thing is they do not want to know." This emphasises not just serious laws in organisational attitude but in the education system that continues to produce young engineers who still think that a career in irrigation is only about building dams and pumping stations. This same argument might also be made about engineers and social science.

Performance-driven organisation

There are many examples of poor organisational performance in the irrigation sector. Some stem from a mismatch between what the organisation was set up to do and what it actually does. The reasons for this are varied and complex. A government research institute may be carrying out research that interests its staff and encourages their career development through publication, but critics may point to drainage practices being promoted that do not fit well with local physical and socio-economic conditions and do not address the priorities of local people.

Typically, there may be poor links between researchers and farmers. Extension services are weakened by a lack of well-trained local professionals and resources to do the job properly. The local professionals have little to offer farmers beyond formalised messages about water management and do not have the skills to cope with today's farmers seeking a livelihood from a range of natural resources. There are many other similar examples.

Incentives play a crucial role in ensuring that individuals and organisations point in the right direction. Salary is important but individuals are motivated by other issues such as career prospects, security of tenure, and the value of working in a worthwhile job. However, there is little point in trying to improve the performance of a research institute by training its staff on the layout of field plot experiments when those same people have to do a second job in the evening in order to secure their livelihood.

Biswas suggests that good organisational performance starts at the top. First and foremost, it is essential to have a good cadre of capable senior managers in place. The institutions and policies, he argues, will then take care of themselves. Browne makes a similar point when he refers to the Indian Administrative Service as a highly-respected organisation working relatively independently of political pressures. It has few problems recruiting capable professionals in spite of the modest pay levels.

Rewards and performance should be more intrinsically linked. In this regard, several international organisations (FAO, IPTRID, ICID and the World Bank) have pooled their efforts to develop new benchmarking methodologies for irrigation system management. The idea is to have a set of indicators and some worldwide records that allow irrigation managers to rank their own performance. Although there are some doubts about the effectiveness of worldwide comparisons, most people agree that, at national level and with comparative systems, the impact of performance benchmarking is one way of boosting the capacity of the management set-up.

Community organisations

The capacity of organisations that function at a community level is vital for individuals to take full advantage of irrigated agriculture. Some are formal but many are informal and not so visible to the observer.

WUAs are the most common formal type of organisation. Although some associations have developed naturally among irrigators as a convenient means of managing common water sources, many have come into being as a result of the current trend of transferring the management of irrigation from state bureaucracies to farmers. There have been successes but there has also been much criticism of the way in which WUAs have been imposed on communities that are not ready for them and that need more time to fully understand and participate in the transfer process. Many farmers face huge difficulties in taking over technical, financial and legal responsibilities for water. The publicity that this has received has perhaps done more than anything else to spur the irrigation community to talk about capacity building.

Rural communities comprise many informal organisations that can have an important influence on their development. This is where social capital lies, and WUAs can fail because of a lack of community capacity or social capital. Community structures can be a very powerful vehicle for capacity development and should not be overlooked.

Empowerment of women

In many communities, women remain excluded from decision-making on water management and agriculture activities. Social development shows that investing in this untapped capital is often a very good option to effectively boost community development and be effective in dealing with food insecurity and poverty alleviation. The engagement of women in water and agricultural life is not only an issue of equity, it is also an issue of efficiency. It is now well documented that women can make a positive difference in the management of WUAs in terms of accounting as well as decision-making.

Private sector

Although government and farmer organisations dominate irrigation in some countries, the private sector, usually in the form of NGOs, is making an impact. An example is the introduction of affordable irrigation technologies for smallholders in Asia and Africa that are manufactured locally and bought by farmers on a commercial basis. This requires organisational structures, known as supply chains, that enable companies to manufacture equipment and sell to farmers through retail outlets or agents. Farmers expect spare parts to be available as well as technical support and credit arrangements in much the same way as any other small-scale enterprise would. The principal prerequisites for this are a strong market for the agricultural produce and profitability for all those in the chain. At present, markets in Asia tend to be distorted by subsidies as governments encourage the development of these new institutional structures. In Africa, they are in their infancy but are more reliant on market forces.

NGOs are well suited to support the creation of these new structures, as they are much better at connecting with people than are government departments. They are also more adaptable to what is essentially a process in which timing cannot be allowed to dictate the outcome.

The Water Sector Level

The sector level is often included as part of the enabling environment. However, this additional level is needed in order to make the point that irrigation is part of the larger picture of integrated water resource management. It reflects the increasing awareness of the need for policies that integrate and cover the whole of the water sector and not just irrigation, water supplies or the environment in isolation.

Enabling Environment Level

This is the most influential level and yet the most difficult level to change. Some issues are well beyond the control of most countries. A typical example is the low international price of basic food crops, which means many farmers are unwilling to grow irrigated crops beyond their immediate family requirements because the production costs exceed the returns. This is fuelled by agriculture policies that favour developed countries and the dumping of surplus food on hungry countries.

However, there are also wider national/local issues to consider that include access to markets, credit and knowledge. Initiatives that set out specifically to make changes at this level are quite rare. They take a long time, in excess of ten years, they are less predictable than contributions at the lower levels, and the expertise available to support them is not readily available either in-country or internationally. However, some initiatives do succeed in making significant changes, and this can be the result of initiatives taken at the lower levels that then come through to change policy. This is often by chance rather than design.

Education and Research

Education and research are issues that concern all levels of the framework from individuals who teach and research, to how they are organised and the kind of environment in which they must function. For this reason, they are here discussed separately from the project-oriented aspects of irrigation development.

The capacity to educate and to undertake research to solve problems is something that all would subscribe to as vitally important to the future of irrigated agriculture. However, few people seem willing and able to take the next step and provide the resources needed.

Education

An educated population is an essential foundation for all aspects of development. Biswas stresses this point, suggesting that "a good education is the foundation of good organisational performance and it starts at the top."

The kind of education that professionals receive must equip them properly for their role. However, in many countries, colleges and universities are poorly resourced. Teachers are poorly paid and have little or no equipment to work with and continue to use outdated curricula that are no longer relevant to the world their students will inhabit. Without institutional changes within the education system, the next generation of professionals and technicians will continue to perpetuate and even reinforce existing outmoded power structures and practices. This also points to the need to develop the capacity of the educators and to provide the resources to produce the professionals and technicians for tomorrow.

Timing is a crucial issue as it takes many years to educate a professional agriculturalist or an engineer. Therefore, both the individuals and the organisations that employ them need to plan well ahead if the demands for such skills are to be met.

In the past 40 years, education and training abroad has been seen as one way of overcoming some of these inadequacies. Governments have sent many of their top professionals for postgraduate training, usually on scholarships provided by the host country or a donor agency. This has exposed young professionals to new ideas and new ways of working, and has served many countries well. However, it is now in rapid decline because of a lack of scholarship funding and the relatively high cost of a course in, say, Europe or the United States of America (US$30 000 for full support for one year) as opposed to the home or a third country (US$2 000-5 000).

Unfortunately, in many instances, this trend towards educating at home or in a third country is not matched by the resources needed to do the job well. The impact in the developed world is also profound as many international training centres have closed or are in decline, with a loss of expertise that will be very difficult to replace.

The rapid changes now taking place in society and technology in the developed world will also affect developing countries. In the developed world, people now expect to pursue a number of career paths in their working lives. This has changed education from a "one-time event" to one of "lifelong learning". This will eventually influence working patterns and education in developing countries.

Research

The capacity to undertake research, like education, is something that all would subscribe to as vitally important to the future of irrigated agriculture. Every developing country has its research institutes. However, they tend to suffer from the same problems that afflict universities, such as inadequate resources, poorly-trained and poorly-motivated staff, and a lack of clear direction as to their contribution to local and national interests.

Funding agencies have not helped this situation. For example, the European Union (EU) usually provides only marginal funding for research and does not recognise overhead costs that national governments are expected to meet. The links between the EU's research directorate and the development directorate are weak, and so there are few connections between research activities and development aid.

The EU has made some attempts to rectify this. However, although the idea was sound, the initiatives have generally failed because of a serious lack of understanding and cooperation between researchers and development agents, mistiming of inputs and poor funding. The funding aspect refers to the process of trying to select the most appropriate research team on the basis of competitive tendering, which usually means on the basis of price. The question often asked of aid administrators in this situation is: would you select your heart surgeon on the basis of price or on the ability to do the job?

There are many lessons here for the development of research organisations and their support so that they can make an effective contribution to national and international interests.

Networks

Networks can have a positive influence at all the levels of

the framework by providing a means of communication for individuals and organisations. The rapid and huge developments in information management provide an enormous opportunity to disseminate and share information. However, at the same time, there is always the threat that a digital gap will benefit the better-off at the expense of the poor and disadvantaged.

The network function of capacity development is two-dimensional. The vertical dimension integrates the levels mentioned in the framework such as the network of people and institutions belonging to a specific project, ranging from farmers to external donors, via local governmental officers and/or training institutes. This line of capacity development was traditionally the one pursued using the top-down approach of projects. It remains active today although with a much more participatory approach and more balance in the decision-making among the different actors.

The horizontal dimension refers to the networks linking actors or organisations of the same nature and level. There is strong evidence that those responsible for development have previously neglected this networking aspect. Social science in agriculture has shown that the adoption of new techniques and the diffusion of progress is often more the result of discussion and comparison among peers than the result of knowledge brought in by other actors.

Linking or twinning a newly-created WUA with an existing one from the same country or from another one is a typical horizontal network. The president of the new WUA is more likely to understand and trust the president of the more experienced one than an external consultant. In the same way, the president is also more likely to adopt a management change suggested by a peer.

The support from aid agencies for this type of networking is growing although it is still on a very small scale. However, the impact usually extends beyond those immediately involved via other networks. There are many examples of this kind of cooperation in the Mediterranean region. This networking is also called professional cooperation. For example, FAO has developed many field programmes based on technical expertise from the "South", known as "South-South" cooperation.

Mixed networks connect actors with similar levels of decision-making but from different fields.

Examples of networks in capacity development

Cap-Net is an international network (horizontal) for capacity building in IWRM and was created in 2001. Its mission is to enhance human resources development for IWRM by means of establishing or strengthening regional capacity building networks.

In France, RNED-HA is a mixed network on water in agriculture that has been very active and instrumental in promoting various technical advances, such as surface irrigation modernisation in the 1980s, and, more recently, best practices for irrigation. The critical advantage of this type of network lies in the fact that it brings together researchers, policy-makers, farmers, industrialists, etc. who each have a say in the decision-making with regard to the applied research programmes.

The Information Techniques for Irrigation System is another example of a mixed network. It was created by the IWMI and the Centre National du Machinisme Agricole, du Genie Rural, des Eaux et des Forets (CEMAGREF) in the 1990s in order to bring together researchers, practitioners and institutional specialists with the goal of developing the capacity to use modern management tools and techniques in irrigation.

The ICID has long been an important international mixed network of national organisations, connecting practitioners, researchers and policy-makers from all over the world, developing and disseminating knowledge, and improving the capacity of its members.

The creation of a network for capacity developers in irrigated agriculture would not only enable people to share information and expertise, it could also become a place for synthesizing experience and lessons learned, identifying research areas, and encouraging the exchange of views. Several countries have already developed their own Web-based networks. FAO has set up a Web site specifically for participatory training in agricultural water management, and includes an e-mail listing to keep members in touch with the latest information.

Networks can be an important tool for practitioners in developing countries. Where people are able to access information and communications technologies, it can reduce the need for expatriate expertise. As Browne puts it, it is a case of "scan globally and re-invent locally".

Diagnosis and Monitoring

The long-term success of designing and implementing capacity building strategies relies to some extent on monitoring the process and the outcomes. However, equally important is having the capacity to do this. Process and inputs are always the easiest parts to monitor. Outcomes and results are usually more difficult to define and assess. However, pre and post-diagnosis should be central to capacity development engineering (CDE).

There are very few examples of monitoring and evaluating capacity development initiatives. They have tended to be very time consuming and expensive, and they can be wide ranging. Benefits can often accrue in unexpected ways that were not obvious at the beginning. They may also arise long after the project has finished. For example, on a project in Zambia the uptake of treadle pumps was not high immediately and on this basis it might be considered to have failed. However, the uptake of other cultural practices was very encouraging as farmers sought to use information in a way that had more immediate benefit for them. This example shows that beyond technical targets, which may or may not be achieved, empowering the actors to take charge of their own destiny can be even more important for capacity development. As mentioned earlier, "empowerment" is a key component of capacity development.

Baselines studies are also needed if sensible measurements of progress are to be made. However, measuring or assessing capacity is not so straightforward.

Lessons for Strategy Development

The issues that come from the framework and several case studies reviewed in the course of preparing this paper provide some interesting and useful insights into capacity development. A list of practical experience and issues can be gleaned from them that might prove useful in the strategy development process.

Policy environment

The importance of a favourable policy environment and political will for development is crucial to successful irrigation development and to developing capacity.

Policies are determined largely by the quality and commitment of the leadership and senior management. Within the policy, capacity development must be guided by clear strategies that address development needs and are nationally owned.

A period of review and experimentation with new approaches is essential before including them in any new policy. Capacity development should be at the heart of development and be pursued on a permanent basis. It is not a temporary fix for correcting investment failures. The approach to capacity development needs to be improved both in terms of concepts and the methods and tools used. There is a need for CDE, with all the features that a corps of engineering implies: accruing knowledge from experience on capacity to diagnose, design, implement, and post evaluate.

Identification and assessment of capacity needs is one of the most difficult tasks. It is a question of who does it and how. It implies strategic planning capacity in a country, and this too may require developing. Experts are needed with interdisciplinary skills, with experience of working at high levels in government or the private sector and who are able to facilitate and not dictate discussion and decision-making. Many countries do not possess such expertise and it may not always be readily available from other countries either.

Full participation of all the stakeholders in the public and private sectors, namely, farmers, engineers, agriculturalists, researchers, educators, politicians and civil servants is essential. Although it is time consuming, it is considered essential to the eventual ownership of any actions taken.

The lack of expertise and the extensive nature of consultation are unlikely to lead to a rapid conclusive or state-of the-art solution. However, it is essential that there be a locally-owned consensus and that it be acted upon. Capacity development is a complex process but complexity must not paralyse it. A framework such as the one proposed in this paper can provide a useful basis for discussions among stakeholders and for pinpointing constraints and priorities.

The challenge is to find ways through the complexity and make useful contributions. As with irrigation engineering, it is important to appreciate the complex nature of irrigated agriculture, but at some point the engineer must be able to interpret this complexity in terms of the relatively simple task of constructing a good irrigation scheme.

The Role of Donors and Technical Cooperation

Although it is accepted that capacity development must come from within, there is also an acceptance that outside help can support and accelerate the process. Partnership is the essence of collaboration between donor and recipient and not the traditional North-South dialogue. This is always difficult to deal with where one is seen to be giving and one is receiving. Establishing true partnerships is a challenge facing both.

Identifying capacity constraints and helping both governments and the private sector to take action can be one of the most effective ways in which external agencies can help. An outside view can often help those inside to focus their attention on the big issues.

The stimulation that external technical cooperation encourages can help to speed up processes and broaden thinking. Long-term collaboration influences the process positively, among others as trust and effective work relationships develop over time.

Some donors may prefer not to become involved in lengthy processes and prefer a project approach. The implications of this in terms of budgets, timing and outputs can be in conflict with the process of capacity development. However, there may be specific actions at the lower levels of organisations, and individuals lend themselves more to the project approach - capacity development projects.

Donors need entry points into capacity development. Typically, the organisation is the level at which donors wish to contribute technical assistance, budgetary and infrastructure support. People selected to help develop capacity need much more than technical knowledge and skills. They should be capable of working in a participatory manner in the wider context of development and be able to convey such ideas to those being trained. This should be reflected in the selection process.

Organisations

Most countries suffer weaknesses at the organisational level. Most organisations supporting irrigated agriculture are government run but there is a growing input from the private sector. It is not a question of either one or the other but of the balance between them. The appropriate strengthening of both must be a priority in any strategy.

A good cadre of capable senior managers that can lead the processes of change is needed if organisations are to develop a good reputation and encourage the best people to join. Salaries are important but not the only criteria. There are many advantages in working with existing organisations rather than inventing new ones. This may not always be attractive but new organisations can be a large drain on already limited management and administrative resources.

Individuals

Individuals are the heart of any organisation and their education and training must reflect the needs of irrigated agriculture and the roles they will play. The "people part of partnerships": partnerships develop between people. While working together, more and better results will be possible over time as mutual trust develops.

Human resources planning offers a tool for predicting the demand from human resources and a logical process for providing them through education and training. It focuses attention on the numbers of individuals needed, their skills, and the education and training infrastructure needed to provide them.

Any strategy must address the young people who are less influenced by rigid organisations, more flexible in their thinking, willing to take risks, and ambitious to build their careers and make a worthwhile contribution.

Education and Research

The capacity to educate future generations in the ways of irrigated agriculture and the capacity to undertake research to solve problems are fundamental to the sustainability of irrigated agriculture. It is essential that both be given a high priority in strategy development.

Education is essential to develop the future cadres of professionals who will become the managers and leaders of organisations and capacity builders. In order to achieve this, the development of capacity within colleges and universities and to some extent in schools will need to be considered.

Research to solve the problems of irrigation and drainage and to encourage the take-up of new ideas may require developing. Research can be most successful, when linked to investment projects, where it can show its value directly.

Sustainability

Ensuring sustainability remains one of main challenges facing capacity development, particularly where it is supported by an external agency.

Sustainability is also a concept that requires careful thought, e.g. when considering whether a training facility should be sustained beyond its usefulness or be closed. Capacity development is not a fixed issue and any strategy must be flexible enough to meet changes in capacity needs.

Points for Discussion

Recognising that a lack of capacity is now perceived as the

main constraint on developing and improving irrigated agriculture, this paper set out: (i) to initiate a debate within the irrigation community on what capacity building means for irrigated agriculture; (ii) to define it as clearly as possible; (iii) to map out the territory it covers and the main issues surrounding it; (iv) to see if it is possible to plan it strategically as is done with more traditional infrastructure projects; and (v) to look for good examples of capacity development from which useful lesson could be learned for the guidance of others.

Many see capacity building (or development) as a rather vague concept. The paper suggests ways of clarifying this concept in the context of irrigated agriculture. It suggests that the UNDP definition provides a good starting-point for understanding capacity development. It embodies the principles of development that are well accepted in the irrigation community.

A simple framework is also proposed that combines and defines the territory of capacity development and irrigated agriculture. It is based on a conceptual approach developed by social scientists and the practical activities of a well-functioning irrigation sector and forms a bridge between them that could provide a useful "picture" of capacity development that is acceptable to irrigation practitioners. It also includes research and education as fundamental activities to capacity development. It is suggested that this "picture" could be a useful tool for participatory discussion between stakeholders and donors for identifying and assessing capacity gaps.

A strategic planning approach is suggested as a means of defining needs and actions. However, careful thought needs to be given to a "project" approach for implementing plans as this may be incompatible with the process approach of capacity development.

A range of basic issues is raised, using the framework as a reference point, which could form an agenda for discussion among stakeholders at the strategy planning stage. In particular, education is stressed as the foundation, a weak one in many countries, on which capacity is built. Thus, it must be a priority on the agenda. The importance of these issues requires confirmation and there may be others that need to be added.

Case studies are used to cover as much as possible of the territory mapped out by the framework to show how capacity can be developed in practice. Although some practical guidance on strategic planning and action can be gleaned, there are serious shortcomings in the literature generally. Most published information tends to be general in nature and to dwell more on infrastructure rather than on capacity development. Those that provide more relevant information tend to discuss individual training issues rather than the larger picture outlined in the framework.

In order for the full nature of capacity development to be taken seriously, it is suggested that more guidance is needed on how it can be done. Much of the conceptual thinking is now complete. What is now needed are more-detailed and carefully-crafted case studies that fully describe the process used to develop capacity. Ways of collecting this information and communicating also need to be devised.

ACCOUNTABILITY MECHANISMS AND TOOLS

A main element of capacity building is the development, introduction and acceptance of accountability mechanisms and tools. In the context of water resource and irrigation and drainage management organisations, three domains of accountability can be identified:

— operational accountability, which involves the mechanisms for monitoring, evaluating, controlling and enforcing the implementation of the service agreement.

— strategic accountability, which relates to the mechanism that allows users to control the formulation services and its costs, for example, by the right to approve or reject working plans and budget proposals.
— constitutional accountability, which relates to the mechanism that allows users to influence the strategic decision-making process by, for example, establishment of assemblies, adoption of statutes and by-laws, and election of representatives in governing and coordinating bodies.

The application of these concepts for the water resources sector in Indonesia will be elaborated further based generic management functions within the organisations and the interaction between different organisations in the sector. It includes the tools that can be developed and introduced in order to enhance the operational, strategic and constitutional accountability related to service delivery and its associated costs and price.

Generic management activities and accountability tools

Traditionally, water resource management and irrigation and drainage agencies have operated in an environment where the level of expenditure on infrastructure and management bears no relation to the water delivery service that they are expected to provide. Until now agencies have seldom defined clearly their level of service and the costs associated with it, mainly because of supply-oriented budgeting systems.

Under the Water Sector Management Programme (WSMP), a service-oriented management approach is adopted that entails an output-driven approach. To achieve this, management needs to know what is an acceptable standard of service and, consequently, the management of resources must be consistent with management goals. A quality assurance system for all activities within the organisations and related out-contracted activities is an effective accountability tool.

A key element in planning is the ability to ascertain the financial and human resources and infrastructure needed to deliver the agreed service based on agreement, transparent processes and accountability, and all subject to a quality assurance system. Clear and unambiguous water rights including their conditionalities are elementary for water allocation, distribution and delivery, and they form an essential tool for operational and strategic accountability on service delivery.

The cost of service provision is related directly to the level of service provided. The higher the level of service, the more management or infrastructure is needed, hence a higher cost is incurred. An asset management plan is a plan for the creation, acquisition, maintenance, operation, replacement, modernisation and disposal of fixed assets in order to provide an agreed level of service in the most cost-effective and sustainable manner.

Asset management plans are effective and transparent tools to determine the present and the future level of costs associated with the agreed level of service. In situations where customers pay the full or partial cost of service provision, the level of service must be balanced against their willingness to pay for the service through a process of consultation. The result of the consultation is then reflected in a service agreement between the organisation and its customers.

A service agreement consists of two main parts: transactions and accountability mechanism to be applied on the transaction. In water resources management the transactions are a certain payment against the delivery of water at an agreed rate, time, place and of a certain quality. The accountability part describes the monitoring procedures to verify whether the transactions are met, the liabilities of both parties in the event that the obligations are not met, what party to address in case of conflict, and what procedure to use to review and update the agreement. Various multitier service relations can be identified: farmer WUA federation of water users associations (WUAF) basin management authority; WUAF district and/or provincial public services; WUAF public and private enterprises and institutes; etc.

Asset management plans also provide information for budget-dependent service providers on the real need for budgeting and the consequences of not meeting budgetary requirements. Cost allocation procedures must be introduced in agencies providing more than one type of service or services to different client or client groups. Cost accounting systems provide a transparent administration and accountability tool to allow service clients to gain insight into the composition of the cost of service they are paying for.

Full or partial cost recovery requires decisions on tariffs and charges. The basis of fixing tariffs, service charges and cross-subsidies must be available in the form of regulations for setting tariffs and charges, and should include the process and procedures involved, including approval by the governing boards. An independent body to prevent political gain at the cost of system sustainability must audit its application. Asset management plans are again an important tool in this process for predicting future expenditures on infrastructure.

Tariffs are often fixed, or the government sets ceilings. Where these are insufficient to cover the actual cost of service provision, subsidies are required to sustain the service provision at the agreed levels. For the WUAF and the International Association of Water Users (WUAI), these subsidies in the form of matching funds are foreseen in the District Irrigation Improvement Funds. The basin management authorities will, for the time being, depend to a larger extent on subsidies. However, in due course, they may also obtain income from charges on bulk water supply, hydrological service provision, and water quality management. Clear subsidy arrangements including rules for eligibility based on performance and urgency of the requested subsidies have to be available. Again, an asset management plan can be an effective tool for indicating the urgency and cost of the required interventions, and the risk and cost in the event of functional failure of the asset.

The service charges can be collected directly from clients and employed for the organisation itself where the organisation has financial autonomy. Clear terms and conditions for payment, including penalties for non-payment, must be made explicit. These in general can be part of the service agreement. Otherwise, charges are collected and reverted into general government revenues. In this case, clear guidelines must exist to guarantee that funds are reallocated to the managing organisation in order to ensure the adequate provision of the agreed level of service.

Budget allocation is to be based on a set of integrated plans, which may include plans for development, asset management plans, human resources development plans, etc. Because strategic plans are dynamic in nature, financial planning usually results in multiyear rolling plans, which may forecast activities for a period ahead, generally 3-5 years. This process involves annual updating based on a review of the previous year's performance and analysis of future needs on service development and associated infrastructural and management adjustments. In case of budget shortfall, asset management plans provide information on the priority of interventions.

Actors in Capacity Building

The introduction of new working relationships and new tasks in the water sector requires major capacity building efforts at the national, provincial, river basin, district and WUA/WUAF/WUAI levels. Given the restrictions on human and financial resources, a multitier capacity building programme is to be developed and implemented. In this multitier approach, the following groups can be distinguished:

The ultimate target groups being those groups that have the tasks and responsibility to manage water resources.

- Temporary (in project) assigned capacity builders who are assigned to train and coach the ultimate target groups in exercising their tasks and responsibilities.
- Temporary (in project) assigned capacity builders and building institutions that are responsible for the training and guidance of the capacity builders above.
- Ultimate postproject capacity builders who will remain available for the ultimate target group to provide on a demand basis continuous training, guidance and assistance after programme completion.
- An independent agency that monitors and certifies capacity builders, and flags distortions to provide timely correction as needed.

Ultimate Target Group

The groups and organisations directly responsible for water management are:

- The WUA/WUAF/WUAI responsible for: (i) water acquisition; (ii) water allocation and distribution among their members; (iii) drainage in their area of jurisdiction; and (iv) maintenance of infrastructure under their jurisdiction.
- The river basin management unit generally responsible for: (i) the day-to-day planning and implementation of water allocation and delivery to the various users in the basin including the WUA-I, based upon an agreed water resources management plan with clear and

explicit water rights; (ii) collection and primary processing of hydrological, meteorological and water quality data; (iii) licensing of abstractions and discharge of effluents; (iv) controlling the destructive power of water (floods, landslides, lahar flows, etc.); and (v) O&M of infrastructure under their jurisdiction.

— The district water resources management service responsible for: (i) guidance of and financial and technical assistance to WUAFs within their area of jurisdiction; (ii) guidance of and assistance to basin authorities within their area of jurisdiction; (iii) O&M of infrastructure under their jurisdiction; and (iv) development of new infrastructure in cooperation with users in their area of jurisdiction.

— Other district services responsible for coordination and planning of activities related to or affected by water and water resource management, such as the planning and coordination service, the environmental monitoring service, and the agriculture service.

— Coordinating bodies at district level, such as: (i) the irrigation commission for coordination and communication among irrigation schemes; (ii) the irrigation forum for consultation and communication among different uses in one irrigation scheme; and (iii) the district-level commission on water resources planning and management.

— Provincial water resources management service for: (i) guidance of and financial and technical assistance to WUAFs within their area of jurisdiction; (ii) guidance of and assistance to basin authorities within their area of jurisdiction; (iii) development and management of the hydrological, meteorological and water quality survey network and processing and dissemination of data for use of long-, medium- and short-term water resources planning and management; (iv) preparation of long- and medium-term river basin development and management plans in basins under their jurisdiction; (v) O&M of infrastructure under their jurisdiction; and (vi) development of new infrastructure in cooperation with users in their area of jurisdiction.

— Provincial-level and basin-level water resources councils responsible for coordination in water resources management (WRM), which includes: (i) establishment of WRM priorities; (ii) regulating WRM including water conservation, water allocation, water use and water hazard control; (iii) regulation of relevant development; and (iv) conflict resolution related to use and management of water resources.

— Central Water Resources Planning Unit (CWRU) responsible for: (i) preparation and development of the long-term national water resources policy and planning in cooperation with other agencies; (ii) development and dissemination of water quality and quantity standards for planning and management of water resources; and (iii) guidance of and assistance to provincial water resources planning units.

— National Hydrological Survey (NHS) is to be established to take care of the development, implementation and guidance of a national hydrological programme. The NHS could be responsible for: (i) planning, development and implementation of national hydrological surveys; (ii) development and dissemination of standards for collection, processing and reporting of hydrological, meteorological and water quality data; (iii) preparation of annual data reports; and (iv) ensuring public access to data for water resource planning and management purposes.

— National Water Council, with a secretariat responsible for coordination and development of the national water resources policy and associated legislation with standing working groups to address national policy, national planning, water sector financing and hydrology/decision-support system data banks.

Postproject Capacity Builders

For a sustainable situation in a development environment, a continuous demand for education, training and support will occur. This demand has to be satisfied by a structural - non-project linked capability in the public and private sectors. These institutions will act on a "commercial" basis to secure their existence. By allowing a variety of institutions to offer education and training services, and by allowing the water resources and irrigation management agencies to select their own programmes and provider, the quality of these services will improve and the cost may fall. Such institutions may also be in-house department training institutes under the Human Resources Development Board with their regional offices, private and public universities and polytechnics, management training institutes, etc.

In addition to the training and education capacity, there is a need for a permanent coaching system for the different management entities, which will allow the introduction of new management, technological and financial systems within the water resource and irrigation management institutions. The establishment of a structural coaching system is encouraged.

REFERENCES

Alaerts, G.J., Blair, T.L., & Hartvelt, F.J.A., "Procedures and partners for capacity building in the water sector," In: *IHE/UNDP*, 1991.

Ayers, R.S., & Westcott, D.W., "Water quality for agriculture," *[FAO] Irrigation and Drainage Paper*, No. 29, 1989.

Easter, K.W., "Economic failure plagues developing countries: en assurance problem," *Water Resources Research*, 29(7): 1913-1922, 1993.

El-Ashry, M.T., Schilfgaarde, I. van, & Schiffman, S., "Salinity for irrigated agriculture," *Journal of Soil and Water Conservation*, 401(1): 48, 1985.

11

Water Management in Dry Lands

In an environment where the natural vegetation is sparse and two normal cropping seasons can pass with virtually no rain, insufficiency of rainfall remains the problem. Conventional farming accelerates loss of soil moisture through reduced ability of the soil to capture, drain and store rainwater. Various examples were shared in relation to PAW, especially situations in which conventional farming was compared with the practice of no till. Cases from Australia, Germany, Kazakhstan, and the United States of America were discussed. All showed PAW increased significantly with no till.

One of the reasons for this effect in the long run is the buildup of organic matter under no-till conditions. An example that demonstrated the difficulties of increasing organic matter in the Sudano-Sahelian Arenosols was presented. Rainwater was often lost as runoff owing to surface sealing because of clay fractions in the surface soil. An example from the Niger demonstrated that the application of calcium (Ca) and potassium (K) fertiliser might improve water infiltration capacity. The addition of organic matter gives stability to sandy soil aggregates, as the sand particles as such do not have any surface forces to bind them.

Despite the economic value of utilising residues in the Sahel, two contributors demonstrated that they are by far the most promising means for improving soil moisture conditions in this zone. Discussion also highlighted that it is necessary to increase crop yield through fertilisation in order to have enough residues to cover all the needs at farm level. Their alternative value as soil stabiliser and as a production input should be better publicised. It is known that biomass production under CA would increase to an extent that the alternative needs for crop residues and the soil fertility needs can be covered. The difficult part is to start this process. Farmers would need a "credit" on biomass.

An alternative to the buildup of organic matter through retaining crop stubble and residues is the application and/ or the incorporation into the immediate topsoil of a mulch of composted organic wastes. Although the application of composted organic waste improves SOM, soil moisture content, bulk density, water infiltration, nutrient distribution and a whole series of soil parameters, it was argued that the effect of ploughing and incorporating fresh crop residue on the biological oxidation of the residual organic matter is similar to "opening the air supply and stirring kindling into the smouldering coals of an old coal stove".

In both cases, the oxidation is accelerated by the improved oxygen supply and accessibility of new readily oxidisable material, resulting in further oxidation of the residual carbon in the system. On the other hand, the retention of crop residues on the soil surface in association with reduced tillage not only reduces erosion, but it also reduces the physical release of carbon dioxide (CO_2) and biological oxidation of soil carbon, which has been a less obvious but usually the greater cause of organic matter depletion in soils.

In addition to these SOM effects, the evaporation from bare soil surfaces and the direct loss through tillage have to be considered. A classic example is the common habit of farmers to wait for rain for seeding, where, in conventional agriculture, the seedbed is prepared shortly before seeding. The often intensive tillage operation associated with seedbed formation dries the soil to tillage depth. Under humid conditions with good soil moisture, the re-compaction of the seedbed at sowing might re-establish the capillary flow of the water to the seed. However, this takes time.

In arid conditions, the seedbed remains dry to the planting depth, so that seeding is only possible after rains that sufficiently moisten the seedbed. This situation can also be problematic in terms of gaining good germination and early plant vigour. Seed is often placed deep to give the seed a chance of staying moist for as long as possible. However, this can lead to slow early growth as roots proliferate quickly only once the plant develops leaves and establishes an energy supply from photosynthesis.

In a no-till, permanent soil cover situation, soil moisture for seedling germination is usually supplied from beneath the seed from stored soil water. Therefore, seeding occurs without the necessity of rain. Moreover, the seeding depth can be shallower, so enabling faster development of the plants and improved resilience. The efficiency of rainfall utilisation in such a system is also higher.

It was emphasized that soil cover is the most important factor influencing water infiltration into the soil. Unlike tilled soils, no-till soils are most receptive to rain and irrigation water as a result of several interacting factors, particularly: the potential for slow infiltration of water through crop residues and enhanced organic-matter-enriched topsoils; the reduction in damaging raindrop impact with subsequent reduction in crust formation; the maintenance of the vertical pore system created by roots, earthworms and other soil animals; improved soil structure and aggregate stability; and improved organic matter mineralisation. These conditions continue to improve with time under no till.

SOIL MOISTURE MONITORING

This applies only in no-till farming systems. In tillage-based systems under arid conditions, soil moisture loss from tillage is considerable, causing the actual supply of water from above to be an important and unpredictable feature for decision-making. Disasters, like floods and landslides, and the costs for road maintenance and water treatment would be minimised where rainwater infiltration and storage in agricultural soils were maximised.

Sufficient evidence for this exists. Moreover, commercial evidence was cited where electricity companies (generating hydroelectricity) have ensured that CA is practised on the lands around the hydrodams. In this way, not only are damaging flash floods prevented but eroded soil does not slowly fill the reservoirs. This was illustrated by the experiences in Paraná, Brazil, where, even in the 1980s, the unexpectedly rapid improvements in catchment hydrology became apparent as a consequence of implementing residue-based no till on significant areas of land/catchments.

Strong climate trends

As the occurrence of extreme climate events seems to be increasing even in moderate climates, the risk in conventional farming is increasing. There is evidence that crops grown under CA suffer less variation in yields in the event of extreme weather conditions. This provides a strong argument for promoting CA as a strategy to adapt to climate change.

Good knowledge of climate patterns and crop requirements enables the development of optimal strategies. Farmers have to deal with several parameters before deciding on a sowing date and crop varieties.

Different methods were suggested (from Bosnia and Herzegovina, Nigeria, and South Africa) for optimising and selecting options. Agrometeorology techniques, which may be based on farmers experiences, and/or simple crop water models give good results, and may be combined with appropriate in-field management. What is expected is a minimum effect of drought on future yield.

The above dilemma and insecurity of farming is a major problem of conventional farming in arid and unpredictable conditions. The aggravation in a conventional system is that the soil is pulverised and transformed into a very fragile condition for preparing the seedbed while at the same time heavy winds or rainstorms happen regularly.

CREATING DROUGHT-RESISTANT SOIL

Restoration of soil porosity

The reason for increased runoff in plough-based systems is that it is most difficult to keep the soil open to capture rain. If the soil is left untouched, it will settle and lose moisture by direct evaporation. Therefore, with conventional farming the soil surface is regularly loosened to break crusts and provide a coarsely aggregated horizon as an evaporation barrier. However, these aggregates tend to disintegrate with every rain. Crust formation was also cited as a major problem in plant-stand establishment in savannah soils in both central and northern Nigeria as well as in northeastern Brazil that were under continuous ploughing. However, the problem disappeared immediately when no till was introduced.

In addition to increasing water infiltration and controlling erosion, soil cover has a major impact in reducing soil temperature, reducing evaporation, increasing available water for plants, enhancing biological activity, contributing to reduce soil compaction and soil crusting, as well as having positive effects on soil chemical, physical and biological properties. All of these are advantageous for the farmer and lead to increased productivity. Furthermore, cropping systems that have permanent soil cover are essential to achieving long-term agricultural sustainability.

Mulch is critical to effective no-tillage systems. This is more difficult to achieve in countries where crop residues are harvested or where long growing seasons and unequal rainfall distribution or wind lessen crop growth. With no-till systems, the soil surface evolves a better-defined structure that results in pores connected continuously to the surface. Mulch will prevent moisture loss through these pores. Equally important, the surface mulch protects the soil surface from the kinetic energy of rainfall and maintains the surface porosity so that infiltration capacity is maximised.

When speaking about the importance of crop residues or roots in the system, it is important to identify what type they are and how they contribute to soil structure, infiltration and nutrient cycling. Different crop types have a range of lignin content and ratios of carbon (C) to nitrogen (N) that can be used to advantage. Moreover, different crops will have different residue geometry and structure. Combined with chemical factors, this will contribute to differential decomposition. Similarly, the differences between annual crop, perennial forage and tree roots are important. They explore different and complementary regions of the soil profile; and they have different decomposition rates that allow for different residence times.

Diverse cropping systems have a net soil benefit as well as other economic benefits. Perennial crops have higher water-use efficiency (WUE) and, because of their deep rooting system, are more drought tolerant. Thus, a combination of trees and crops may optimise the system. WUE also increases with fertility levels. Nutrient cycling can be enhanced through aboveground and belowground residue management planning.

Runoff farming and water harvesting

Water harvesting is a very old practice. It is gaining more and more attention in western Sudan, and results are very encouraging. In dealing with agricultural production in western Sudan (semi-desert to semi-arid climates), new technologies need to take into account the socio-economic aspects of the area. One example presented reported that farmers used to sow and weed in one operation. This practice led to the loss of valuable amounts of moisture at the beginning of the growing season before seeds could be sown. In short, this meant that runoff was harvestod when there was no crop, and the crop was grown when there was no runoff to harvest. An integration of technologies was required.

Runoff farming is site specific. However, while importance is given to the diversion of rain-runoff water into reservoirs, very little is done about the unabated soil erosion in catchment areas. This erosion results in a buildup of sediment concentration in the runoff water, which in turn results in the siltation of these reservoirs. The outcome is a very short life span for these reservoirs.

Other methods of rainwater harvesting at a small scale were discussed, e.g. demi lunes (Niger) and zai pits (Mali and Niger). The demi lunes are 3 m x 3 m at their widest part and are connected with each other by earth bunds producing a continuous water-harvesting structure. The assumption is that farmers will maintain these structures because they will benefit from the increased fruit yields that water harvesting will give. The demi lunes not only harvest water but also the fine soil particles loaded with organic material that "fertilise" the trees continuously. Spectacular growth, beyond expectations, was generated through improvements to the traditional planting pits in Mali,

Niger, etc. The so-called zai/tassa improved planting pits catch more of the sparse rainfall. They also concentrate organic matter in the form of dung/compost into these pits, and the plant nutrients it contained enable the plants to make efficient use of the rainwater that is captured more effectively. "Improved water infiltration" could be subdivided into both the rate of true infiltration and the quantity of water entering the soil.

Runoff farming and water harvesting are noteworthy. One alternative to prevent sedimentation problems is the use of agroforestry in catchment areas where cattle grazing is forbidden or restricted. It was recognised that this last suggestion is the most difficult to implement in the field.

Soil moisture management and organic matter decomposition

Soil is a combination of inert material, roots and soil biota and it is in the soil that the transformation begins of not-living environmental substances to living ones, which is expressed by soil fertility. The use of cover crops changes the competition and dominant species distribution in the soil and plant cover and life cycles of plants and insects involved in plant pollination.

Effective microbes (EM) can increase porosity by increasing the favourable organisms in the soil, and some experimental results seem to show that avocado trees need less water when these organisms are present. When crop residues are left undisturbed on top of the soil, a kind of "soil skin" is generated. This "skin" normally provides food and shelter to both microlevel and mesolevel flora and fauna, thereby increasing the soil biotic load and biological activity. In addition, macrofauna contributes to organic matter transport and long-term storage in deep soil layers.

Although the principles that lead to good soil moisture management are not new, farmers do not apply them widely, mainly because adapted technologies are not developed by researchers

It was suggested that in many places there has been little effort by researchers to visit field sites to monitor and measure the effects on soil conditions (biological, chemical, physical and hydric) of techniques that farmers themselves have devised and/or developed, put into practice, and found to be worth continuing. This is particularly evident in the case of soil moisture conditions.

Most effects of treatments are judged on the basis of yield alone even though the expression of yield is the result of many different resource-interactions that occur over time within both the soil and the plant. There is considerable information available on responses to nutrient ions, but much less is known about how such responses are affected by the sufficiency or insufficiency of soil moisture, and by interactions with organic materials and processes in soil.

The question was raised: "Is this scarcity of in-field investigations on soil moisture causes/effects because this sort of muddy-boots work does not represent 'whizz-kid' science, or because of other reasons for being unwilling to leave the research stations?" At present, research seems to be following and explaining why some farmers' practices work well.

Research should also promote integrated and appropriate techniques. Any such information could add to that which researchers already have, or could also be the spark for new lines of research work. It also provides an opportunity for farmers and researchers alike to see the principles in action in real-life situations.

Four requirements for the adoption of good land management practices were cited:

— For any new technology to be adopted successfully by farmers, it must bring the farmer a visible and immediate benefit - economic or otherwise.

— This benefit must be substantial enough to convince the farmer to change existing practices.

— For the technology to be disseminated on its own, the costs incurred in its adoption must be affordable for the farmer.

— The introduction of the new technology should be followed up by an extension service for a long period.

Farmers themselves will be the major source of solutions once they are engaged in the development process. The problems cannot be resolved without working with farmers in real field situations. As one of the contributors stated: "the reason why technologies are not adopted by farmers is very often simple, at least for the people who are working directly with the farmers: the issue and these techniques have not been thought and conceived with the farmers but by technicians or scientists studying the problem only from their discipline and their technical point of view."

A systemic approach that can integrate all the parameters could be the solution to this problem. Cultural and socio-economic knowledge and an excellent capacity for understanding and exchanging with farmers are fundamental to the sharing of concepts and practices. In addition, in-field experimentation is required in order to observe the interaction of the different components and variables of the system.

The major barriers are often not agronomic but a complex mix of governance, policy, economic and institutional deficiencies. The successful adoption of no till in Argentina and neighbouring countries has in great measure to be attributed to the common sense of their farmers and their ability to detect the economic, physical and other advantages of the system, and how it fits well with general conditions and development bottlenecks.

In addition to being more cost-effective and environmentally friendly, no-tillage systems are no more complex or challenging than other methods of crop production. The complexity tends to be in the minds of researchers and advisers who are unable or unwilling to consider the multitude of factors that contribute to the emergence of a problem and, hence, its possible solutions.

A large component of the problem of limited technological change among resource-poor farmers arises from inefficient research and extension systems that utilise a linear model of information and knowledge flow. Impact has been small because the resource-poor farmer is not integrated into efficient knowledge or information systems.

Another problem associated with resource-poor farmers is their aversion to risk. Their livelihoods and those of their families depend on the decisions they take, and one bad decision can lead to hunger, farm loss or worse. CA is less risky than conventional, tillage-based agriculture with respect to most of the common environmental hazards that affect agriculture. However, innovative farmers do need some support during the early stages. They do not need paternalism and free inputs, but just some assurance that someone will look after them if things go "horribly wrong". With CA, things seldom go "horribly wrong" because it is far more knowledge intensive than it is technically intensive.

Changing from ploughing to direct seeding is not a major undertaking, but changing the whole system is difficult to carry out, involves a lot of stakeholders at the same time, and needs to be understood by the farmers. Furthermore, the socio-economic environment for farmers is often not beneficial. Critical to achieving acceptance and widespread uptake of CA are the establishment of long-term integrated partnerships between farmers, researchers, agribusinesses, engineers, financiers and local politicians, and the combining of new technologies and farmers' initiatives with zero financial risk to the farmers.

One major issue with no-till farming is that it has to be continuous before it can be effective both in terms of soil quality improvement and yield increase. It is not inevitable that yield decreases in the first years after changing to no till. Different reasons and management solutions were discussed to prevent yield losses. These included level of technology, full plant stand, and placed fertilisers, etc. that lead to increased rainwater productivity, and stable and even better yields.

The discussion highlighted the fact that the knowledge and technology for no-till systems are not always available in developing countries. Thus, the development of systems that require less management and knowledge and which allow the crop residues to remain on the soil surface until the time for the next cropping season might be more useful in such areas. In this way, moisture is conserved and the soil surface is protected from rain and wind impact.

Moreover, owing to the absence of crust as the result of the residue cover, more rainwater will infiltrate into the soil and less will run off. At the time of planting, a one-off cultivation prepares the bed for planting, eliminates most weeds, and incorporates the residues into the soil, so providing the conditions for some organic matter buildup and additional nutrient in the soil. However, the key issue is probably to understand the necessity and the reason to change and evolve.

Limiting Runoff and Water Infiltration

In a given watershed area (village level), improved agricultural practices cause water to infiltrate slowly into the subsoil and aquifers of the catchment. This water regenerates in surface wells and in ponds/tanks downstream. Results are positive and many: runoff is negligible; soil erosion is negligible; agroforestry becomes well established (grazing forbidden); and water is available in wells and tanks for domestic and farming purposes. This changes the whole socio-economics of the village.

IMPROVED SOIL MOISTURE MANAGEMENT

Local Soil Management Practices

This statement generated contrasting views. The discussion highlighted the significant downstream effects of local soil moisture management practices. Thinking in larger dimensions of space and time, it could be suggested that such practices could have a marginal effect on the rising of the sea level. Other contributors indicated that the effects (cleaner surface water and reduced sedimentation of lakes and dams) may be expected more at the regional than at the global level.

Carbon sequestration, especially under tropical conditions, was considered by some contributors to be of minor importance. Optimisation of green water (on-site) management improves rainwater-use efficiency and thus yield potential. At the same time, it improves blue water (off-site) resources, in particular through reduced runoff - hence, fewer flash floods and reduced erosion and water turbidity. Thus, it increases groundwater recharge and creates a more stable river baseflow. However, a recent study concluded that downstream impacts of soil management practices on hydrological regime would be felt more readily at small scales.

Off-site effects are less clear and the impacts are site-specific and dependent on many factors. As one participant stated: "We should be careful when generalising that off-site, or "blue water", impacts of soil conservation are always significant, and positive. Rather, we need to evaluate carefully where positive land-water linkages exist from the downstream users' perspectives that merit an investment on their part."

One such evaluation was shared during the conference. It concerned dairy farms in KwaZulu-Natal, South Africa. The most pertinent aspect was a report from Umgeni Water, in whose catchment area these farms lie. Each year, water quality is tested in addition to ongoing water flow measurements. The latest annual report states that the water sampled was of the highest quality measured anywhere in KwaZulu-Natal and the best recorded in 40 years of testing. In addition, water production had increased with minimal silt loads. This is after only about four years of no-till farming. This report is important because it contains probably the best justification for large-scale adoption of CA practices, emphasizing additional environmental benefits of this farming method: "This, in addition to carbon sequestration benefits, will surely help greatly in convincing several governmental departments of the need to seriously consider adopting CA as the national standard for agricultural production."

One participant made the point that where the impact of soil water conservation on downstream hydrology is small, then at the watershed-scale "badlands" or degraded lands that are not suitable for agriculture may contribute much of the sediment. However, as another participant asked: to what extent does a systematic approach to soil moisture management on a large scale actually influence water regimes downstream? In parts of India where the pressure on land and water is great, there is concern that watershed management approaches focusing on a better use of water at field or microwatershed level reduce the amount of water available downstream.

It was suggested that conventional systems and no-till systems should be compared with the natural system that existed in the watershed before humans began to place large demands on the soil and water resources. Moreover, perhaps we should imagine what the landscape looked like 200 years ago under native, perennial vegetation and wonder whether a mature no-till system looks more like that older system. In every case when this comparison is made, optimised no-till systems are clearly superior to any system that has tillage involved. Natural systems cycle water through macropores in a soil that is covered with litter. No-till systems do the same. Tilled systems have water running off the soil into the streams. Carbon does not decline in healthy natural systems and proper no-till systems, but it does decline in tilled systems.

As one contributor stated: "In many cases, human pressure on the resources (soil and water) is too great. In the case of competition between soil and animals for crop residues, if we put such pressure on natural ecosystems, they would also be degraded rapidly. This is exactly where the problem is. There are large areas in semi-arid regions where population pressure is probably too great for a natural ecosystem to survive. The question is then: What can we propose to the rural populations in those regions and what are the implications for agriculture and soil moisture management?"

Production of Biomass for Soil Cover

The costs of biomass as "production input" are easily offset by the cost reduction in land preparation and downstream damage. If soil conservation measures are adopted on a large part of the watershed area, increased evapotranspiration resulting from the maintenance of permanent soil cover could reduce water availability at the outlet of the watershed. This effect may be offset by the improved infiltration and aquifer recharge. However, this depends very much on the rainfall and the conservation measures adopted. Put simply, annual rainfall of less than 400 mm causes little or no recharge, while rainfall of more than 800 mm (particularly with large intense events) causes significant recharge. Between these two limits, recharge will only take place in areas where localised runoff ponds.

Soil Moisture Management Practices

Where rain falls on a dry soil, the air in the soil pores is trapped beneath a downward-moving wetting front.

Where the air cannot escape, the weight of water causes pressure buildup in the soil air until it is able to support the weight of the water, and thus slow down and halt further downward percolation. Such a problem of obstructed infiltration is worse in a de-structured soil than in one that has wormholes, old root-channels and large porous soil aggregates.

Only life-processes in soil can create such structures as continuous macropores, which create matrix areas of different levels of moisture and aeration within the soil. These facilitate soil life processes that affect nutrient dynamics. Nutrient cycling can be enhanced through aboveground and belowground residue-management planning. Diverse cropping systems have a net soil benefit as well as other economic benefits. Different crop types have a range of lignin and other organic products that can be used to advantage. Different crops will have different residue geometry and structure. Combined with chemical factors, this will contribute to differential decomposition.

Relatively uniform moisture conditions in the soil are important for several reasons. The population and diversity of biota will be enhanced, resulting in improved nutrient availability for plants. In addition, crops are more tolerant to salinity, even where it is high, compared with the high levels of fluctuation in salt levels under flood irrigation as the soil moves from saturation at field capacity to varying degrees of desiccation. Therefore, improved soil moisture management, especially combined with the use of organic fertilisers, can be a very effective means of overcoming micronutrient deficiencies. Reduced tillage and diverse cropping systems utilising different rooting morphology, perennial crops, etc. can also stop or mitigate salinisation. Where adoption of no till reduces the incidence or duration of runoff and flooding to depressional landforms, then less methane (CH_4) is generated - a good thing for greenhouse-gas mitigation.

The reason for mass movement of saturated soil from slopes was put forward as a result of the substitution of deep-rooted trees with shallow-rooting grasses. One case presented illustrated that good practices can be harmful in regions where landslides are a prevalent phenomenon. An example was presented where, in order to avoid landslides, the infiltration rate was reduced to a minimum, and crude oil byproducts and other materials were used to reduce the infiltration rate and soil moisture storage.

One intervention discussed the disturbance imposed by driving over the soil with wheels carrying loads of more than 1 tonne (or perhaps less). A study found that this had a devastating effect on earthworms and all soil macrofauna. Microfauna effects were more varied.

TOOLS AND TECHNOLOGIES FOR SOIL MOISTURE MANAGEMENT

The modelling of the effects of reduced runoff (by improved soil and water conservation techniques) on the water-balance components (crop transpiration, soil evaporation and deep percolation) may assist in the understanding of the process and contribute to practical questions for the whole watershed, which includes downstream effects. Validation of model-simulation results by groundtruthing is necessary with present available indicators such as groundwater level and river discharges.

Available Equipment and Techniques

There is a lack of available equipment and techniques for simple, routine, low-cost measurement and monitoring of soil water. Tools in the form of equipment to support soil moisture monitoring are usually inaccessible to most poor farmers. It is important to develop low-technology field-useful approaches that give farmers and land users simple field-applied techniques and models for determining the onset and severity of drought. Preferably, methods based on the experience of the farmers themselves in estimating watering needs should be developed. These methods should be based on the observation of plants and on empirical evaluation of the humidity by touching soil samples.

One such method is the measurement of evaporation. Compared with the calculation of potential evapotranspiration, it is very simple, acceptable by farmers and includes meteorological conditions (solar radiation, relative humidity, air temperature and wind). Based on the measured evaporation, farmers know exactly the quantity and time to irrigate their crop. Another method is the use of "indicator plants" in order to avoid relying on instrumentation altogether.

The purpose is to provide a visible warning of impending drought stress within a commercial crop. In addition, it is important to consider plant roots as their distribution better integrates the soil volume than do most installed field sensors. One system of providing a simple, cheap, field-usable measure of soil hydraulic conductivity (soil water infiltration) that does not require trained people to measure or simulate and process the information is now available. It is called the Visual Soil Field Assessment Tool (VS-FAST).

The method is simple and robust, being based on fundamental, globally tested and accepted soil physical principles. This ensures that the operator (seen principally as being farmers and local extension staff in developing countries) can use the technique after basic training. The soil probe, mentioned in the discussion paper "Adequate tools and technologies to support soil moisture management", is useful for determining stored soil moisture in order to make decisions on continuous cropping and what crop type to pick. Dr Paul Brown of Montana State University developed the probe in the 1960s. It gained popularity in the 1970s and 1980s as farming practices in the Northern Great Plains in the United States of America decreased the use of a summer fallow year in order to conserve soil moisture.

A major feature of the probe is the inclusion of a piece of a wood auger below the ball on the tip. At any point of insertion of the probe the handle can be given a quick twist and the probe removed. A small sample of the soil at the depth of insertion can be brought to surface and the degree of soil moisture can be determined by feel and appearance. Similarly, where there is more resistance to the probe insertion, a small sample of soil can be brought up in order to verify that it is a dry layer.

For some years now, many Australian irrigation farmers have successfully used capacitance probes to monitor soil water content. The probes are installed, calibrated and maintained; and the data analysed and interpreted for the farmer by local commercial companies.

Estimation of Plant-available Soil Water

"Breaking down reality" - the reductionist approach - does not necessarily reveal or anticipate how actual combinations of factors will affect crop development.

A new family of models called artificial neural networks (ANNs), which are neither reductionist nor mechanistic, have the potential to be useful in the management of soil moisture. The ANN methods embody to some degree the non-linearities that characterise the combined effect of variously timed water applications. The notion of relating water inputs directly to yield may seem unreasonably simplistic to some. However, for irrigation planning purposes at least, there appears to be some value in applying a non-linear algorithm in a straightforward manner.

Another approach is the "least limiting water range" (LLWR) - this being the water content range within which plant growth is least limited. The strength of this approach is the integration of three physical factors affecting plant growth directly, but its use still is limited by difficulties of some soil related property measurement. As it is conceptually well based, it should be used as a simple procedure by farmers and field technicians to evaluate soil quality and propose mitigation management techniques in order to overcome the soil physical constraint factors.

Plant physiology should also be taken into account when water availability to plants is assessed. This is because the efficiency with which the plant uses water can vary considerably.

Soil Moisture Monitoring

The available methods for soil moisture measurement all need trained people to apply them and they are not acceptable to or usable by small farmers. Some of them are very expensive and need laboratory measurements and field calibration. In addition, the results of measurements by these methods are not precisely exact. Moreover, they need accompanying methodologies for translating the results into a soil management strategy. Many provide only an approximation or a trend until properly and fully calibrated or cross-checked against crop variables.

Measured Concepts

Process-based models are difficult to design. There are issues of selection of variables and functional forms, and they need testing with observed data. It is extremely difficult to include interactions between factors in models. Any model is a simplification of the reality. However, from the scientific point of view, choosing a model is also selecting a priori phenomena that will be more important than others for describing, exploring or explaining a situation. Based on the example from the working document and the reality faced by farmers in Brazil, it was suggested that "different kinds of farmers may decide to go or not to go for residue application for entirely different reasons and, so it seems, on perfectly rational grounds." Therefore, a coherent analytical framework, clearly stating for what circumstances the observations and experiences are valid, is needed so that a greater focus can be achieved.

A holistic approach to soil moisture management is needed. While studying a system functioning, and in order to find the proper explanation for a given outcome, it is normally necessary to examine the area of interactions and to understand and quantify them.

Short-term Benefits and Long-term Externalities

There is a need to concentrate on identifying natural soil properties that lead to some soils providing more moisture to crops particularly at mid-season droughts or at the end of a generally short growing season. In addition, the identification of management practices that optimise soil moisture supply to crops is vital, which must depend on local conditions of climate, soil and crop. It is the scientific understanding of how such practices actually operate that will enable:

— generalisation into general principles;
— upscaling to larger areas;
— identification of alternative options that can vary with local circumstances and availability of resources.

ADOPTION OF DROUGHT-PROOFING PRACTICES

Technology Adoption

Farmers have to be involved in the development of new technologies from the outset. The more changes required of farming systems, the more essential farmer involvement becomes. Most technologies on soil drought-proofing require a profound change in farming practices, not just the change of a component as with improved varieties or fertilisers. This means that farmers have to learn how to integrate new practices into their systems. They will have to pay for this learning with setbacks in the first few years.

Successful adaptation usually takes some time. Pioneers in a community are viewed critically and risk becoming outsiders. This is especially valid for Africa, where strong farmers associations are lacking. Added to this is insufficient access to information, inputs, credit, etc. Support programmes are often of too short a duration, e.g. 2 years, followed by programmes with new foci.

Farmers want advice on how to improve their way of farming and how to become more profitable. They do not want technological packages as the interactions of the components are difficult to understand, nor do they want to feel that they are being used as guinea pigs by researchers and extensionists. Researchers and extensionists are often too output oriented, forgetting the production costs and the risk of not covering even production costs when adequate markets and prices are lacking.

Implementation of Soil Moisture Management Systems

The word subsidy is loaded with different interpretations in different countries and different circumstances. In most industrialised countries, agriculture subsidies consist of input and price subsidies. These subsidies have had devastating impacts on the components of agricultural systems through soil erosion, water pollution, loss of biodiversity and genetic erosion, loss of habitats and buffer zones, etc. and on farmers. Recently, in Europe and elsewhere, subsidies have also been applied as a payment for environmental services of environment-friendly agricultural practices.

Although subsidies and legislation can be an obstacle to the adoption of sound soil management practices, they can also become a very effective tool for their promotion when they are used to support the development of sustainable on-farm soil management investment. "Moderate" subsidies can become a useful incentive for farmers to change existing habits. For example, in the Netherlands, the Government assisted the large-scale introduction of greenhouses using a well-designed subsidy strategy. If this kind of subsidy was "right" for a developed European country, why is it "wrong" for developing countries? The issue is to what point subsidies will prevent producers from improving their practices because they make as much money with the subsidies as they would by improving their production.

Farmers do not feel they need to change anything. Moreover, those who are willing to change find it difficult to adopt crop rotations that do not include the subsidised crops. Apart from industrialised countries, few governments now pay subsidies or provide other incentives to farmers. Governmental support of the agriculture sector in the United States of America was mentioned as an example during the discussion: "If the majority of farmers in the United States of America were playing by the real-world market rules, as the majority of the South American farmers are, they would go out of business unless they adopted technological changes to reduce costs, like using CA extensively or, directly, no till."

Appropriate Soil Moisture Management

A variety of factors can contribute to the adoption or rejection (and, hence, failure or success) of technically perfect technologies. While economics is not the least of these, other factors are also important, such as labour requirements, different priorities, and local sensitivities.

It was argued that among all factors influencing the adoption rate of soil moisture management techniques, the economic factor ranks far ahead of considerations such as education, training and information. The simple fact that it is difficult to maintain or replicate pilot experiments shows that, even when a technology is known, other factors must play an important role. The adoption of CA by farmers in the states of Paraná and Santa Catarina in Brazil undoubtedly contributed to their being able to survive the combined effects of low output prices and increasing input prices in the 1990s. It was found that the most important reasons for not adopting soil fertility options in the Sahel were:

— farmers lack capital, and credit systems are poorly developed and relatively inaccessible;

— support is in many cases required during the transition period as soil depletion is financially far more attractive for farmers than soil improvement;

— creation of an enabling environment on a regional or national level requires important investments.

Microeconomists would say that farmers are perfectly rational and maximise some utility function. It should be well understood that this maximum utility consists of different attributes, whose mix varies with each farmer. Farmers know the advantages and disadvantages of their production systems very well.

Consequently, they will often be reluctant to adopt technologies that deviate drastically from a system that they know well and with which they have considerable experience with respect to its performance under variable weather conditions. For example, the farmers in the Shiwaliks Belt of the Himalayan ecosystem are illiterate and bound by tradition.

It has been observed that, being illiterate, they are not aware of various programmes of research and extension being carried out by different agencies. As such, they are not in a position to know about emerging technologies. Second, being bound by tradition, the farmers are not prepared to forgo the existing practice and adopt new ones. Farmers should be involved while carrying out research on a particular problem. In this way, they will learn about the new technique and will consider it their own invention. The major obstacles identified in southwest Nigeria were:

— Solutions not relevant. Farmers perceived them as unnecessary because of tradition and old methods.
— Traditional and cautious behaviour of farmers. Existing knowledge attitude and practices were not initially considered carefully by researchers.
— Insufficiently qualified and motivated extension services.
— Communication plan. Research results were not communicated properly and adequately to end users.

In this case, the role of a scientist should be to develop of a drought-resistant variety of black gram rather than insist on the adoption of green gram. Extension staff can appear not motivated when, in fact, owing to a lack of training, they are fearful or ignorant. They need to have the training and practice. Then they will gain in confidence and be successful. Farmers are reluctant because they are comfortable with what they have always done and they "fear" failure with adopting a new system.

Moreover, they usually have the equipment to continue doing what they have always done and they are either unable or unwilling to invest in the equipment necessary for a change in the production system. This confirms the view that farmers who are comfortable with their current situation would not look for a change, especially if it is an unknown and, hence, risky one. A number of conditions necessary for adoption can be summarised:

— the technology should result in an immediate and significant improvement in farmers' profits;
— the benefit/cost ratio should be high;
— the investment (monetary and labour) should be within the fiscal and physical capacity and within the existing work habits and traditions of the farmers;

— there should be frequent follow-up;
— an integrated (holistic) participatory approach should be taken;
— credit should be available;
— a well-designed marketing strategy is necessary.

Feed for cattle in arid regions

This is not valid for most dry regions and it is a significant problem that needs to be addressed. Livestock feed is generally in short supply, and the use of crop residues is the easiest way to feed livestock. In the short term, there are some farmers, particularly in dryland areas of developing countries, who do not have a choice and need to use crop residues for animal feed. In most parts of Africa, the right of land use is limited to the growing season. Fields are open for communal grazing after harvest. It is difficult for individual farmers to prevent livestock entering their fields. This is also valid for cover crops, which are commonly grazed by freely roaming cattle or goats.

This is also the case in the dairy industry in Uruguay, which has a completely different climate (temperate subhumid) from the semi-arid areas of Africa. In dairy farms, almost all the aerial biomass is grazed, harvested for hay or silage, or trampled by the animals. Nevertheless, the crops of these systems (oats in winter, and corn or sorghum in summer) are in rotation with perennial pastures that are also grazed. However, they leave a very important amount of roots in the surface ploughed horizon and develop progressively a mulch of residues on the surface in the 3-4 years that they occupy in the rotation.

In the Shiwaliks region, farmers are reluctant to use crop residues for mulching as they have inadequate sources of fodder for their cattle. Farmers prefer to use locally available mulch materials such as the leaves of some local trees for moisture conservation or they adopt other techniques for this e.g. breaking capillaries in order to check evaporation losses. Here again, the aim should be to develop the existing technology of using locally available mulch materials.

A primary role for the international breeding institutions is the selection of varieties that produce more straw. In general, conservation agriculture is perceived as more profitable by farmers, while researchers and extensionists perceive it as more risky. Only about 3 percent of the land in the world is being farmed using CA principles.

As one participant stated: "If farmers truly perceived CA as more profitable, I cannot believe that more of them would not be adopters. There are many reasons for not adopting, but tradition and fear are among the main ones. My perception is that they do not see enough upside potential for them to make a change. They would if they really perceived more profit in changing."

Another participant stated that the reason for the lack of uptake is that farmers usually do not know of CA and its potential. The critical mass of this knowledge has not yet been reached. In addition, there are forces working against it, e.g. the oil industry and the agricultural machinery industry, that have paramount importance and power in developed countries. It is not difficult to imagine that they would not be happy with a technological change that reduces oil use and replaces the use of huge machinery with a few machines and medium-sized tractors.

REFERENCES

Anderson, J.M. and Ingram, J.S.I., "Tropical Soil Biology and Fertility," *A Handbook of Methods*, 2nd Edition, CABI, Wallingford, 1993.

Balpande, S.S., Puranik, R.B. and Joshi, R.P., "Nutrient potential of crop residues under sorghum-pigeon pea intercropping in vertisol," *Annals of Plant Physiology*, 5, 259-262, 1991.

Bationo, A., Christianson, C.B. and Mokwunye, U., "Soil fertility management of the pearl millet-producing sandy soils of Sahelian

West Africa: the Niger experience," In: *Soil, Crop and Water Management in the Sudano-Sahelian Zone,* Proceedings of an international workshop. ICRISAT Sahelian Center, Niamey, Niger. January 11-16, 1987. ICRISAT, Patanchera. 159-168, 1989.

Fryrear, D.W., "Wind erosion: mechanics, measurement and control," In: *Soil, Crop and Water Management in the Sudano-Sahelian Zone,* Proceedings of an international workshop, ICRISAT Sahelian Center, Niamey, Niger, January 11-16, 1987.

Klaij, M.C. and Hoogmoed, W.B., "Crop response to tillage practices in a Sahelian soil," In: *Soil, Crop and Water Management in the Sudano-Sahelian Zone,* Proceedings of an international workshop, ICRISAT Sahelian Center, Niamey, Niger, January 11-16, 1987, ICRISAT, Patancheru, 265-275, 1989.

Lewis, A.H., Procter, J. and Hood, A.E.M., "A comparison of ley and arable farming systems," *Journal of Agricultural Science* (Cambridge), 54, 310-317, 1960.

Powell, D.L., "The management of irrigated saline and solodic soils", In: *Russell's Soil Conditions and Plant Growth,* (Edited: A. Wild), 11th Edition, Longman Scientific and Technical, Harlow, 927-951, 1988.

12

Water Conservation Measures

IRRIGATION PERFORMANCE INDICATORS

Numerous indicators, usually called efficiencies, have been developed to assess irrigation performance. The definitions of efficiencies as formulated by the ICID/ILRI are widely used. According to the ICID/ILRI, the movement of water through an irrigation system involves three separate operations: conveyance, distribution, and field application. In congruence, the main efficiencies are defined for water use in each of these three operations.

The conveyance efficiency (e_c) is the efficiency of the canal or conduit networks from the source to the offtakes of the distribution system. The e_c can be expressed as the volume of water delivered to the distribution system plus the non-irrigation deliveries from the conveyance system divided by the volume diverted or pumped from a source of water plus inflow of other sources. The distribution efficiency (e_d) is the efficiency of the water distribution (tertiary and quaternary) canals and conduits supplying water to individual fields.

The *ed* can be expressed as the volume of water furnished to the fields plus non-irrigation deliveries divided by the volume of water delivered to the distribution system. Finally, the water application efficiency (e_a) is the relationship between the quantity of water applied at the field inlet and the quantity of water needed for evapotranspiration by the crops to avoid undesirable water stress in the plants. Unlike the ASCE Task Committee on Describing Irrigation Efficiency and Uniformity, the ICID/ILRI do not make a further distinction between consumptive, beneficial and reasonable water use.

Source reduction aims to reduce the volume of drainage water by improving the irrigation performance. In addition to the efficiencies defined by the ICID/ILRI, this report uses a number of performance indicators defined by the ASCE Task Committee on Describing Irrigation Efficiency and Uniformity to assess the appropriateness of water use in an irrigation system and its contribution to the production of drainage water.

The ASCE Task Committee performance indicators are defined by the water entering and leaving the boundaries of a system. To judge the performance of an irrigation system, the ASCE Task Committee groups the fractions of water leaving the system boundaries in a specified time period into categories of use: consumptive versus non-consumptive use; beneficial versus non-beneficial use; and reasonable versus unreasonable use. This analysis requires an accurate description of the components of the hydrologic balance within defined boundaries of a control volume over a specific period of time. The time period taken is a cropping season rather than a single irrigation event.

Consumptive uses include water that finds its way into the atmosphere through evaporation and transpiration, and water that leaves the boundaries in harvested plant tissues. Non-consumptive use is water that leaves the boundaries in the form of runoff, deep percolation and canal spills. The partitioning of applied water into beneficial and non-beneficial uses is a distinction between water consumed in order to achieve agronomic objectives and water that does not contribute to this objective. Beneficial use of water supports the production of crops and includes evapotranspiration, water needed for improving or maintaining soil productivity, seed bed preparation and seed germination, frost prevention, etc. Non-beneficial uses include deep percolation in excess of the leaching requirement, tailwater, evapotranspiration by weeds, canal seepage, spills, etc.

Not all non-beneficial uses can be avoided at all times. Due to physical, economic or managerial constraints and various environmental requirements, some degree of non-beneficial use is reasonable. For example, reasonable but non-beneficial deep percolation can occur because of uncertainties that farmers face when deciding how much to irrigate to replenish the available soil moisture. Unreasonable uses include those that do not have any economical, practical or environmental justification.

Once defined, the water balance components can be used to make rational decisions about the appropriateness of the water use and whether they have a positive or negative impact on crop production, economy, regional hydrology, and on the amount of drainage water.

Three of the performance indicators proposed by the ASCE Task Committee are important in relation to the planning and design of conservation measures. The first indicator is the irrigation consumptive use coefficient (ICUC) which is expressed as follows:

$$\text{ICUC} = \frac{\text{Volume of irrigation water consumptively used}}{\text{Volume of irrigation water applied} - \Delta\ \text{storage of irrigation water}} * 100\% \qquad (1)$$

The *ICUC* deals with the fraction of water that is actually consumed in the system beneficially and non-beneficially. The denominator contains 'change in storage of irrigation water' or that part of the water applied that has not left the control volume and is thus usable during succeeding periods. The quantity (100-*ICUC*) represents the percentage of non-consumptive use of the applied water. Thus, the *ICUC* deals with the fraction of water that is irrecoverable and therefore cannot be re-applied elsewhere.

The second indicator, irrigation efficiency (*IE*) deals with water used beneficially for crop production. *IE* can be expressed as:

$$IE = \frac{\text{Volume of irrigation water beneficially used}}{\text{Volume of irrigation water applied} - \Delta \text{ storage of irrigation water}} * 100\% \qquad (2)$$

The quantity (100-*IE*) represents the percentage of non-beneficial uses of applied water.

$$IS = \frac{\text{Volume of irrigation water beneficially and / or reasonably used}}{\text{Volume of irrigation water applied} - \Delta \text{ storage of irrigation water}} * 100\% \qquad (3)$$

The ASCE Task Committee suggests a third performance indicator because not all water losses are avoidable. The irrigation sagacity (*IS*), first introduced by Solomon, is now defined by the ASCE Task Committee as:

The ASCE Task Committee suggests the use of *IS* to complement *IE*.

The numerical values of these three performance indicators provide appraisals on the overall effectiveness of the irrigation system and its management and the contribution towards drainage water production. However, the use of these performance indicators relies on quantification of the various water uses and fate pathways of water. There are many methods for determining the volume associated with each water use. Some are direct measurements, some are indirect measurements and some are obtained as the closure term from mass balance. Each method has errors associated with it that affects the accuracy of the performance indicator. Clemmens and Burt have shown that the confidence interval for the *ICUC* is about 7 percent.

The accuracy of *IE* is in general less than the *ICUC*, as quantifying the beneficial water use is normally quite difficult. Therefore, they concluded that reporting more than two significant numbers for performance indicators is inappropriate. Furthermore, they introduced an approach using component variances to determine the relative importance of the accuracy of the variables that contribute to the estimate of the performance indicator. The component variances can be used to determine which measured volumes need closer attention. Improving the accuracy of the components with the highest variances will have the greatest impact on improving the accuracy of the performance measures.

SOURCE REDUCTION THROUGH SOUND IRRIGATION MANAGEMENT

Source reduction aims to reduce the drainage effluent through improving irrigation performance and thus improving *IE*. As certain losses are unavoidable, source reduction aims more precisely to increase *IS* by reducing non-beneficial, unreasonable uses.

Reasonable Losses

This section explores the various losses captured by subsurface field drains and open collector and main drains, and establishes estimates for reasonable losses.

Losses captured by subsurface field drains

The discharge of on-farm subsurface drainage in irrigated agriculture may be determined with the following water balance equation:

$$q = \frac{R + S_C + S_I + S_V - D_m}{t} \quad (4)$$

where:

q = specific discharge (mm/d);

R = estimated deep percolation (mm);

S_c = seepage from canal (mm);

S_l = lateral seepage inflow to the area (mm);

S_v = vertical seepage inflow (mm);

D_{rn} = natural groundwater drainage from the area (mm); and

t = time period of measurement or calculation (d).

Deep percolation losses

Table 1 shows that each irrigation method has a range of inherent distribution uniformity, e_a, and deep percolation. The data reported are based on well-designed and well-managed systems on appropriate soil types.

Table 1. Estimated reasonable deep percolation losses as related to irrigation methods

Application method	*Distribution uniformity (%)*	*Water application efficiency (%)*		*Estimated deep percolation (%)*
		Tanji and Hanson, 1990	*SJVDIP, 1999f*	
Sprinkler				
- periodic move	70-80	65-80	70-80	15-25
- continuous move	70-90	75-85	80-90	10-15
- solid set	90-95	85-90	70-80	5-10
Drip/trickle	80-90	75-90	80-90	5-20
Surface				
- furrow	80-90	60-90	70-85	5-25
- border	70-85	65-80	70-85	10-20
- basin	90-95	75-90		5-20

In Table 2, e_a is defined as the ratio of the average amount of water stored in the rootzone to the average amount of water applied. Deep percolation, surface runoff, tailwater and evaporation losses make up the total losses. The e_a reported by Tanji and Hanson is based on data estimated from irrigated agriculture in California, the United States of America, where there is a state mandate to conserve water and where most irrigators receive some training.

More recently, the Technical Committee on Source Reduction for the San Joaquin Valley Drainage Implementation Program reported e_a values that differ somewhat from the 1990 values. The 1999 values were based on an analysis of nearly 1 000 irrigation system evaluations and they represent updated practical potential maximum irrigation efficiencies.

In contrast, previously published FAO application efficiencies for irrigation application methods and e_a values reported by the ICID/ILRI are substantially lower, representing a lower level of management and probably a less than optimal design. A properly designed system that is well managed can attain quite high efficiencies. On heavy soils, surface and drip irrigation can attain similar levels of efficiency.

Estimates for deep percolation have been made on the basis of the following assumptions: no surface runoff occurs under drip and sprinkler irrigation; during daytime sprinkler irrigation evaporation losses can be up to 10 percent and during night irrigation 5 percent; tailwater in furrow and border irrigation can be up to 10 percent and evaporation losses up to 5 percent; and no runoff occurs in basin irrigation and evaporation losses can be up to 5 percent.

Thus, reasonable deep percolation losses may vary with the irrigation application method and water management employed.

Canal seepage

Canal seepage varies with: the nature of the canal lining; hydraulic conductivity; the hydraulic gradient between the canal and the surrounding land; resistance layer at the canal perimeter; water depth; flow velocity; and sediment load. The canal seepage can be calculated using empirically developed formulae or solutions derived from analytical approaches. Canal seepage might also be estimated on the basis of Table 2.

Table 2. Seepage losses in percentage of the canal flow

Type of canal	*Seepage losses (%)*
Unlined canals	20-30
Lined canals	15-20
Unlined large laterals	15-20
Lined large laterals and unlined small laterals	10-15
Small lined laterals	10
Pipelines	0

Excessive seepage can occur due to poor canal maintenance. Any seepage in excess of the aforementioned figures needs to be regarded as unreasonable.

Seepage inflow

Seepage inflow from outside areas is discussed in the section on land retirement. Discussion of seepage inflow from deep aquifers requires substantial geohydrological knowledge, which is beyond the scope of this publication.

Losses captured by collector and main drains

The drainage discharge in collector and main drains depends on the discharge of field drains and additional inputs such as irrigation runoff from farmers' fields, and operational and canal spills. The ILRI in collaboration with the ICID analysed irrigation efficiencies for 91 irrigated areas on the basis of questionnaires submitted by 29 national committees of the ICID. For the interpretation of the data, the basic climate and socio-economic conditions were taken as primary variables.

On the basis of these variables, the irrigated areas were divided into four groups. Group I includes all areas with severe rainfall deficit, and the farms are generally small with cereals as the main crop. For Group II, the main crop is rice and the rainfall deficit is less than for Group I. Group III has a shorter irrigation season than the first two groups and the economic development is more advanced. In addition to cereals, the most important crops are fodder crops, fruits and vegetables. For Group IV, irrigation is supplementary as this group has a cool, temperate climate.

The operational or management losses in the conveyance system are related to: the size of the irrigation scheme; and the level of irrigation management, communication systems and control structures, i.e. manual versus automatic control. The size of the tertiary or rotational units also has a significant influence on the operational losses. Bos and Nugteren estimate that optimum efficiency can be attained if the size of the rotational unit lies between 70 and 300 ha. Where the rotational units are smaller, safety margins above the actual amounts of water required are introduced, as the system cannot cope with temporary deficits. Larger rotational units require a long filling time in relation to the periods that the canals are empty, as the canals are relatively long and of large dimensions. This requires organisational measures to correct timing, which is often difficult.

In Egypt, canal tail losses are estimated to account for 25-50 percent of the total water losses in irrigation. For Egypt, it is expected that operational losses can be reduced significantly when measures such as automatic controls and night storage are introduced. Other losses reaching collector and main drains are from the distribution system.

In addition to the seepage losses from the tertiary and quaternary canals, the method of water distribution, farm size, soil type and duration of the delivery period affect the e_d.

Management Options for On-farm Source Reduction

Improving on-farm irrigation management

Improving on-farm irrigation management involves optimising irrigation scheduling. This means determining when to irrigate and how much water to apply. This can be done by using real-time weather data to obtain the reference crop evapotranspiration (*ETo*) and taking the product of *ETo* and crop coefficient to establish the depth of water to apply. Soil water balance methods are commonly used to determine timing and depth of future irrigations. The California Irrigation Management Information System (CIMIS) and FAO's CROPWAT are based on these principles.

Basic assumptions in these methods are that healthy high-yielding crops are grown and that the soil moisture depletion between two irrigations equals the crop evapotranspiration. The latter assumption may not be valid where the plant roots are extracting water from shallow groundwater. Where such conditions exist, using CIMIS or CROPWAT will give an overestimation of the depletion and result in more water being applied than needed to replenish soil moisture.

Field experiments have been implemented all over the world to evaluate the contribution of capillary rise from a shallow water table towards crop water requirements. However, no simple calculation procedures have been developed yet. Ayars and Hutmacher propose a modified crop coefficient to incorporate groundwater contribution to crop water use. A manual calculation procedure is proposed in the section on shallow water table management. Rough estimates for capillary rise might be used, especially where drought resistant crops are grown or during periods when the crop is less sensitive to water stress. FAO has determined the yield response to water for a range of crops. Such information may be helpful in assessing the risks involved, in terms of yield losses, when rough estimates of capillary rise are used in optimising irrigation scheduling.

Soil sampling and soil water sensing devices can provide valid estimates of soil moisture depletion. Provided these instruments are properly installed and calibrated, and the users adequately trained, irrigation scheduling can be based on the results.

Improving water application uniformity and efficiency

Uniformity and e_a can be improved by upgrading the existing on-farm irrigation system or by converting to a technique with a potential for higher efficiency and uniformity. Surface irrigation by gravity flow is the most common irrigation technique as it does not involve the costs for the O&M of pressurised systems. For this reason, over the coming decades surface irrigation is likely to remain the predominant approach, although better uniformity and e_a can be obtained with sprinkler and drip irrigation.

There are several ways of improving the performance of furrow irrigation. The first method is to reduce the furrow length. This measure is most effective in reducing deep percolation below the rootzone for field lengths exceeding 300 m. The set time has to be reduced at the same time, and is equal to the difference in the initial advance time and the new advance time. Failure to do so will greatly increase the surface runoff and subsurface drainage. A second option is to apply cutback irrigation. Cutback irrigation means reducing the inflow rate of irrigated furrows after the completion of advance.

A third option is to use surge irrigation. Surge irrigation is intermittent application of water to an irrigation furrow. Initial infiltration rates in a dry furrow are high. As the water continues to run, the infiltration rate reduces to a constant rate. If water is shut off and allowed to infiltrate, surface soil particles consolidate and form a partial seal in the furrow, which substantially reduces the infiltration rate. When the water inflow into the furrow is re-introduced, more water moves down the furrow in the previously wetted area and less infiltration into the soil takes place. This process is repeated several times. As the previously wetted part of the furrow has a lower infiltration rate and the advance in this part is higher, the final result is a more uniform infiltration pattern.

Last, better land grading and compaction of the furrow can improve the uniformity and efficiency of furrow irrigation. Improvements in basin irrigation consist mainly of adjusting the size of the basins in accordance with the land slope, the soil type and the available stream size. FAO gives guidelines on how to estimate optimal basin sizes. To obtain a uniformly wetted rootzone, the surface of the basin must be level and the irrigation water must be applied rapidly.

Drip and sprinkler irrigation systems have the potential to be highly efficient. However, where the systems are not properly designed, operated or maintained, the efficiency can be as low as in surface irrigation systems. Improving the uniformity and thus the efficiency in drip and sprinkler irrigation involves reducing the hydraulic losses. Losses can be minimised by selecting the proper length of the laterals and pipeline diameter and by applying appropriate pressure regulation throughout the system. After the system has been properly designed and installed, good O&M of the system is crucial. Phocaides provides guidelines on the design and O&M of pressurised irrigation techniques.

For additional details on on-farm irrigation management, the reader may refer to publications by Hoffman et al., and Skaggs and Van Schilfgaarde.

Options for Source Reduction at Scheme Level

At a scheme level, numerous options can be applied to reduce the conveyance, distribution and operational losses. To reduce seepage and leakage, canal rehabilitation or upgrading might be required. To reduce the operational losses, improvements in the irrigation infrastructure and communication system could be implemented. To increase the distribution efficiency, tertiary and quaternary canals might need to be upgraded.

Moreover, an increase in delivery period might help to increase the distribution efficiency. Furthermore, a number of policy options are available. In many countries, driven largely by financial constraints, the water users now manage the irrigation systems. It is hoped that handing over the systems to the water users will raise efficiency and profitability. FAO has developed guidelines for the transfer of irrigation management services.

Impact of Source Reduction on Long-term Rootzone Salinity

The main objective of source reduction in the context of drainage water management is to reduce the amount of drainage water. For the reduction of the amount of subsurface drainage this means that the amount of water percolating below the rootzone will be reduced through improving water application efficiency. In areas where salinisation is a major concern, it is important to assess the feasibility and impact of source reduction on rootzone salinity.

The equilibrium rootzone salinity is a function of the salinity of the applied water that mixes with the soil solution and the fraction of water percolating from the soil solution. This can be expressed as:

$$EC_{f_rR} = EC_{SW} = \frac{EC_{IWi}}{LF_i} \tag{5}$$

where:

EC_{frR} = salinity of the percolation water which has been mixed with the soil solution (dS/m);

EC_{SW} = salinity of the soil water (dS/m);

EC_{IWi} = salinity of the infiltrated water that mixes with the soil solution (dS/m); and

LF_i = leaching fraction of infiltrated water that mixes with the soil solution (-).

Therefore, it is important to evaluate whether the average net amount of percolation water under proposed irrigation practices satisfies the minimum leaching requirements to avoid soil salinisation. In this evaluation, rainfall should be considered as it might supply adequate leaching. The leaching fraction of the infiltrated water that mixes with the soil solution can be expressed as:

$$LF_i = \frac{f_r R^*}{f_i I_i + P_e} \tag{6}$$

where:

f_r = leaching efficiency coefficient of the percolation water (-);

f_i = leaching efficiency coefficient related to the incoming irrigation water that mixes with the soil solution (-);

I_i = irrigation water infiltrated, which is the total applied irrigation water minus the evaporation losses and surface runoff (mm);

P_e = effective precipitation (mm); and

R^* = net deep percolation (mm).

In this equation, it is assumed that over the irrigation seasons a shallow zone of water is created below the rootzone which has a salinity equivalent to the percolation water. Where this assumption is not correct or where deficit irrigation is actively practised under shallow groundwater conditions, capillary rise and deep percolation have to be entered as separate entities in the rootzone salt balance. In this case, the soil water salinity in the rootzone is a function of the salinity of the infiltrated water, capillary rise and percolation water.

It is often assumed that the salinity of the deep percolation water is equivalent to the average rootzone salinity. However, due to irrigation and rootwater extraction patterns, the salinity is lower in the upper portion of the rootzone due to higher leaching fractions (zone of salt leaching) and the salinity is higher in the bottom portions because of smaller leaching fractions (zone of salt accumulation). Under normal irrigation and root distribution, the typical extraction pattern for the rootzone is 40-30-20-10 percent water uptake from the upper to the lower quarter of the rootzone. Equation 10 can be used to calculate the rootzone salinity of five successive depths under this water uptake pattern to obtain finally the average salinity in the rootzone.

Maintaining a Favourable Salt Balance under Source Reduction

The relationship between EC_{IWi} and the average soil salinity of the saturated soil paste (EC_e) can be calculated for each LF_i and expressed as a concentration fraction (Table 1). These concentration factors can be used to calculate the relationship between EC_e and EC_{IWi}. Where the salinity of the infiltrated water and the crop tolerance to salinity are known. If the y-axis of the figure were the threshold salinity for the crop under consideration (EC_{ts}), then the diagonal lines would give a range of leaching requirements (*LR*) expressed as a *LF*.

Various researchers have developed other methods to calculate the LR to maintain a favourable salt balance in the rootzone. An empirical formula developed by Rhoades and Rhoades and Merrill is:

$$LR = \frac{EC_{IW}}{5 * EC_{ts} - EC_{IW}} \quad (7)$$

where EC_{ts} is the threshold salinity for a crop above which the yield begins to decline and EC_{IW} is the salinity of the infiltrated water. The value 5 was obtained empirically.

The salt equilibrium equation (Equation 7) can be used to assess the feasibility and impact of source reduction on the rootzone salinity by calculating the *LR* expressed as *LF*, and to compare the results with the expected percolation under improved irrigation practices. The amount of percolation water should cover the *LR*. A safety margin is advisable because irrigation uniformity is never complete in the field. Depending on the distribution uniformity of the application method, the actual *LR* should be 1.1-1.3 times higher than the calculated *LR*. For example, for drip, sprinkler (solid set) and basin irrigation, smaller safety margins might be adopted while for border irrigation a larger margin might be more appropriate (Table 2).

Impact of Source Reduction on Salinity of Rootzone

General information

The drainage pilot study area is situated in the south east of the Punjab, Pakistan. The area has been suffering from waterlogging and salinity problems for a long time and therefore was selected as a priority area in urgent need of drainage. The main causes of waterlogging are overirrigation and seepage inflow from surrounding irrigated areas and canals. The estimated average seepage inflow is 0.5 mm/d with a salinity of 5 dS/m. In 1998, a subsurface drainage system was installed on a pilot area of 110 ha. The design discharge of the system is 1.5 mm/d and the design water table depth is 1.4 m. The soils are predominantly silty (q_{fc} 0.36) with an estimated leaching efficiency coefficient (f_i) of 0.9. The climate is semi-arid with an annual potential crop evapotranspiration of 1 303 mm and an average annual effective rainfall of 197 mm.

The main crops are wheat in winter (December to April) and cotton in summer (June to October/November). Both crops are irrigated throughout the year, as rainfall is insufficient to meet crop water requirements. Irrigation water is supplied through open canals and has a salinity of 1 dS/m. Wheat is irrigated through basin irrigation and cotton is irrigated with furrow basins.

The major application losses occur as a result of deep percolation due to poor field levelling (non-uniformity) and as a result of uncertainties in relation to rainfall and water distribution. The annual average e_a is 64 percent $((ET_{crop}-P_e)/I)$, including 5 percent evaporation losses. The depth of the rootzone is assumed to be 1 m and the water uptake pattern is 40, 30, 20, 10 percent from the first to the fourth quarter of the rootzone, respectively. It is assumed that no capillary rise occurs.

Long-term rootzone salinity

Long-term salinity in the rootzone under these conditions is calculated using Equations 1-11 presented in Annex 4 for the four-layer concept. The first step is to calculate *ECIWi* using Equation 10:

$$EC_{IWi} = f_i I_i/(f_i I_i + P_e) * EC_I = (0.9 * 0.95 * 1718)/(0.9 * 0.95 * 1718 + 197) * 1 = 0.88 \text{ dS/m}$$

The second step is to calculate the LF_i values for the infiltrated water mixing with the soil solution for the successive quarters:

$$LF_{i1} = (IW_i - 0.4 * ET_{crop})/IW_i = (0.9 * 0.95 * 1\,718 + 197 - 0.4 * 1\,303)/(0.9 * 0.95 * 1\,718 + 197) = 0.69$$

$$LF_{i2} = (IW_i - (0.4 + 0.3) * ET_{crop})/IW_i = (1\,666 - 0.7 * 1\,303)/1\,666 = 0.45$$

$$LF_{i3} = (IW_i - (0.4 + 0.3 + 0.2) * ET_{crop})/IW_i = (1\,666 - 0.9 * 1303)/1\,666 = 0.30$$

$$LF_{i4} = (IW_i - (0.4 + 0.3 + 0.2 + 0.1) * ET_{crop})/IW_i = (1\,666 - 1.0 * 1\,303)/1\,666 = 0.22$$

The third step is to calculate the *ECe* values with Equation 7. Considering that for most soils EC_e » 0.5 EC_{sw}:

$$EC_{e1} = 0.5\ EC_{sw1} = 0.5 * 0.88/0.69 = 0.64\ dS/m$$
$$EC_{e2} = 0.5\ EC_{sw2} = 0.5 * 0.88/0.45 = 0.98\ dS/m$$
$$EC_{e3} = 0.5\ EC_{sw3} = 0.5 * 0.88/0.30 = 1.47\ dS/m$$
$$EC_{e4} = 0.5\ EC_{sw4} = 0.5 * 0.88/0.22 = 2.00\ dS/m$$

The average EC_e of the rootzone, including the first value of the infiltrated water, is 1.11 dS/m.

Source reduction and the impact on rootzone salinity

Disposal of drainage water in southern Punjab is a major problem. Only very limited volumes of drainage water can be disposed into main irrigation canals and rivers. Evaporation ponds have been constructed to relieve the disposal problem but adverse environmental effects are observed. To sustain irrigated agriculture and to prevent serious environmental degradation, parallel conservation measures have to be implemented. As the calculated rootzone salinity of 1.11 dS/m is far below the threshold salinity value of wheat and cotton, source reduction at field level should be considered. Source reduction might be attained through a combination of measures including: precision land levelling; the shaping of basin and basin furrows; improved irrigation water distribution; and the introduction of water fees.

The threshold EC_e value for wheat is 6 dS/m and for cotton 7.7 dS/m. To assess the minimum leaching requirement, the threshold rootzone salinity of the most sensitive crop in the crop rotation will be used. Theoretically, the leaching fraction of the percolation water mixing with the soil solution could be less than 5 percent. If the LF_i value at the bottom of the rootzone is assumed to be 0.05, the total irrigation water applied can be calculated as:

$$LF_i = (f_i I_i + P_e - ET_{crop})/(f_i I_i + P_e)\ 0.05 =$$
$$(0.95 * 0.9 * I + 197 - 1\ 303)/(0.95 * 0.9 * I + 197)\ 0.05$$
$$* (0.855 * I + 197) = (0.855 * I - 1\ 106)$$
$$I = 1\ 374 \text{ mm}$$

In this case, the e_a is 80 percent, which is within the range of reasonable deep percolation losses. Higher efficiencies cannot be achieved as farmers will need to cope with uncertainties in rainfall and to a certain extent in water supply. Second, some losses always occur as a result of non-uniformity and farmers' inability to apply exact amounts of water. Equation 10 (Annex 4) can be used again to calculate the salinity of the infiltrated water mixed with the soil water:

$$EC_{IWi} = (0.9 * 0.95 * 1\ 374)/$$
$$(0.9 * 0.95 * 1\ 374 + 197) * 1 = 0.86 \text{ dS/m}$$

The LF values for the infiltrated water mixing with the soil solution and the EC_e values for the successive quarters are calculated the same way and result in an average rootzone salinity of 2.77 dS/m against the original value of 1.11 dS/m. Indeed, the salinity increased, but it is still acceptable for the common crops grown in the area. The above calculations were based on dividing the rootzone into four equal parts (quadrants). However, this model can be extended into n-number of layers provided that the rootwater extraction pattern is known, e.g. in 15-cm depth increments for 90-cm rooting depth.

Impact of Source Reduction on Salt Storage

In previous sections, long-term steady-state conditions were assumed to prevail. However, salinity levels during a cropping season are not stable and change throughout the season. To study the impact of irrigation and drainage measures on crop performance, it is important to know the changes in rootzone salinity during a cropping season over multiple time periods. These periods may range from a day to a monthly period. The mass of salts at the start of a period and at the end of a period normally differs and can be expressed as:

$$S_{start} = S_{end} - DS \tag{8}$$

where:

S_{start} = quantity of salts in the rootzone at the start of the period (ECmm);

S_{end} = quantity of salts in the rootzone at the end of the period (ECmm); and

DS = change in salt storage in the rootzone (ECmm).

The salt storage equation is defined as:

$$\Delta S = \frac{IWEC_{IW} - \dfrac{R * S_{start}}{W_{fC}}}{1 + \dfrac{R*}{2W_{fC}}} \tag{9}$$

where:

IW = infiltrated water (mm); and

Wfc = moisture content at field capacity (mm).

Equation 9 can be used to calculate changes in soil salinity within a cropping season. The salt storage equation can also be applied to the four-layer concept. Using the data from the previous example, the following calculation example assesses the impact of source reduction on the rootzone salinity on a monthly basis.

Impact of Source Reduction on Salt Balance of the Rootzone

To explore the changes in salinity over the growing season as a result of source reduction, the salt storage equation (Equation 9) applied to the four-layer concept as explained in Annex 4 is used. For each layer the change in salt storage is calculated using Equation 22 as presented in Annex 4 for the first layer and Equation 23 for the consecutive three layers. The average EC_e is based on the quantity of salts stored in the rootzone at the end of the calculation periods.

Impact of Source Reduction on Salinity of Drainage Water

To estimate the impact of source reduction on the extent of reuse and disposal of drainage water, it is necessary to consider not only the water quantity but also the changes in water quality. Generally, the salinity of the drainage water increases as the drainage discharge decreases. Where all the deep percolation is intercepted by the subsurface drainage system, the amount of generated drainage water is:

$$D_{ra} = f_r R^* + (1 - f_r) R^* + (S_i - D_m) \qquad (10)$$

The salt load of the subsurface drainage water is the salinity of the intercepted water multiplied by the depth of drainage water plus the salt load of the seepage inflow minus the salt load of the natural drainage. If the natural drainage can be ignored the salt load of the subsurface drainage salinity can be calculated as follows:

$$D_{ra} EC_{Dra} = f_r R^* ECfr_R + (1 - f_r) R^* EC_1 + (S_i EC_{S_i}) \qquad (11)$$

where:

EC_{Dra} = salinity of the subsurface drainage water (dS/m);

EC_{frR} = salinity of the percolation water that mixed with soil solution (dS/m); and

EC_{Si} = salinity of the seepage inflow intercepted by the subsurface drains (dS/m).

EC = salinity of the irrigation water (dS/m)

The salinity of the subsurface drainage water (EC_{Dra}) is the salt load as calculated with Equation 11 divided by the depth of drainage water.

In the California studies, there appears to be some correlation between salinity, boron and selenium in terms of the waste discharge load, i.e. changes in flow result in similar changes in loads of salts, selenium and boron. This load-flow relationship exists because the shallow groundwater contains excessive levels of salinity, selenium and boron. Where this relationship exists in an area, regulating the salinity load would also regulate boron and selenium loads in drainage discharge.

The salinity of the subsurface drainage water is diluted in the main drainage system by surface runoff and canal spills. These losses will be drastically reduced under irrigation improvement projects. For example, in Egypt it is expected that through the introduction of night storage reservoirs, the canal tailwater losses will be reduced to almost zero.

Impact on Drainage Water Generation and Salinity

Source reduction influences not only the rootzone salinity but also the amount of drainage water generated and the salinity of the drainage water. This is of special interest when seeking options for reuse and disposal. Under normal conditions and source reduction, the amount of generated drainage water in the drainage pilot study area in Pakistan can be calculated using Equation 10:

$$D_{ra}\text{-normal} = 363 + 163 + 182.5 = 708.5 \text{ mm}$$

$$D_{ra}\text{-reduction} = 69 + 131 + 182.5 = 382.5 \text{ mm}$$

Thus, when source reduction is applied, 46 percent less drainage water is produced. A major reduction is obtained in the first months of the winter season and to a lesser extent during the summer season.

The salinity of the drainage water increases mainly during the first two months of the winter season when source reduction is applied in comparison to the normal irrigation practices. The increased rootzone salinity that develops over the summer season as a result of reduced leaching and the subsequent leaching at the start of the winter season are the main reasons for this increase in salinity of the drainage water.

The total salt load is normally of interest for disposal options. The salt load can be expressed as the depth of water multiplied by the salinity of the drainage water (ECmm). The annual salt load is calculated using Equation 11:

$$D_{ra}EC_{Dra}\text{-normal} = 363 * 2 * 2.0 + 163 * 1 + 182.5 * 5 = 2\ 528 \text{ ECmm}$$

$$D_{ra}EC_{Dra}\text{-reduction} = 69 * 2 * 8.60 + 131 * 1 + 182.5 * 5 = 2\ 230 \text{ ECmm}$$

This means that the annual salt load reduction achieved under source reduction is 11.8 percent. If the *TDS* (g/litre) is approximately 0.64 *EC* (dS/m), the annual reduction is salt load is 1.9 tonnes per ha.

Shallow Water Table Management

In the past, drainage systems were typically designed to remove deep percolation from irrigated land plus any seepage inflow. In saline groundwater areas the depth and spacing was also determined to minimise potential capillary rise of saline groundwater into the rootzone. However, the interaction between crop water use from shallow groundwater and irrigation management has been used to demonstrate the potential impact on irrigation design and the active management of drainage systems. However, there has been little research on the active management of drainage systems in arid and semi-arid areas. The concept of controlled subsurface drainage was first developed for humid climates.

Controlled Subsurface Drainage

The concept of controlled subsurface drainage can be applied as a means to reduce the quantity of drainage effluent. The size of the areas where the water table is controlled by one structure depends on the topography. During system operation, it is important that the water table be maintained at a relatively uniform depth. A detailed topographical map is essential for dividing an area into zones of control for the water management system. Butterfly valves could be installed to restrict the flow from individual lateral lines, and manholes with weir structures could be installed along the collector. In situations where the drain laterals run parallel to the surface slope, in-field controls might be necessary to maintain uniform groundwater depths.

Raising the outlet or closing the valves at predetermined times maintains the water table at a shallow depth. Withholding irrigation applications at the same time induces capillary rise into the rootzone. In this way, plants meet part of their evapotranspiration needs directly from soil water, replacing irrigation water with shallow groundwater that would otherwise have been evacuated by the drainage system. Without irrigation applications, the water table gradually drops in areas with no seepage inflow from outside, while a more or less constant water table can be maintained in areas with high seepage inflow.

Where shallow water table management through controlled drainage is practised, the drainage effluent and salt load discharged is reduced. Controlled drainage also helps to reduce the loss of nutrients and other pollutants in subsurface drainage water. Maintaining a high groundwater table significantly decreases the nitrate concentration in the drainage water. This decrease is a result of a load reduction in drainage discharge and an increase in denitrification.

Considerations in Shallow Water Table Management

In arid to semi-arid areas the main purpose of drainage is the control of waterlogging and salinity. Maintaining the water table at a shallow depth and thus inducing capillary rise into the rootzone seems counterproductive to attaining this objective. However, projects in Pakistan and India have shown that the salts, which accumulated in the rootzone when shallow groundwater was used for evapotranspiration, were easily leached before the next cropping season. In monsoon-type climates, the accumulated salts are leached in the subsequent monsoon season. This experience shows that shallow water table management is an important mechanism for reducing the drainage effluent and at the same time for saving water for other beneficial uses.

The extent to which shallow water table management can be used to reduce the drainage effluent discharge depends on:

- capillary rise into the rootzone, which is related to the water table depth, soil type and water table recharge;
- salt accumulation in relation to the salt tolerance of the crops; and
- potential to maintain a favourable salt balance.

Capillary Rise

Capillary flow depends on: soil type; soil moisture depletion in the rootzone; depth to the water table; and recharge. Evapotranspiration depletes the soil moisture content in the rootzone. Where no recharge through irrigation or rainfall takes place, a difference in potential induces capillary rise from the groundwater. In the unsaturated rootzone, the matrix head, which is caused by the interaction between the soil matrix and water, is negative. At the water table, atmospheric pressure exists and therefore the matrix head is zero. The water moves from locations with a higher potential to locations with lower potentials. Capillary rise from the groundwater or underlying soil layers to the rootzone takes place under the influence of this head difference.

Darcy's Law can be applied to calculate the capillary flow. However, the unsaturated hydraulic conductivity, *K(q)*, needs to be known. A difficulty is that *K(q)* is a function of the moisture content and the moisture content is a function of the pressure head.

Although water movement in the unsaturated zone is in reality unsteady, calculations can be simplified by assuming steady-state flow during a certain period of time. The steady-state flow equation can be written as:

$$z_2 = z_1 + \left[\frac{h_2 - h_1}{-1 - \frac{q}{K(\bar{\theta})}} \right] \qquad (12)$$

where:

$z_{1,2}$ = vertical coordinates (positive upward and z = 0 at groundwater table) (m);

$h_{1,2}$ = soil pressure neads (m); and

K(q) = unsaturated hydraulic conductivity (m/d).

$K(\bar{\theta})$ is the hydraulic conductivity for $\bar{h}$ which is $(h_1 + h_2)/2$.

With this equation, the soil pressure head profiles for stationary capillary rise fluxes can be calculated. From these pressure head profiles, the contribution of the capillary rise under shallow groundwater table management can be estimated.

Instead of calculating the capillary rise manually, computer programs might be used. Programs are available based on both steady-state and non-steady-state models. An example of a steady-state model is CAPSEV. Non-steady-state models include SWAP. SWAP is not specifically designed to calculate capillary rise but mainly to simulate water and solute transport in the unsaturated zone and plant growth.

Favourable Salt Balance under Shallow Water Table Management

The salt balance and salt storage equations apply for situations where the downward flux dominates over the capillary rise. When shallow groundwater table management is implemented, capillary rise is induced and the upward flux dominates over the downward flux. Where no irrigation and percolation occurs during the calculation period, the change in salt storage in the rootzone equals:

$$DS = GEC_{gw} \tag{13}$$

The change in salt storage is a function of the amount of capillary rise and the salinity of the groundwater. The accumulated salts in the rootzone need to be leached before the next growing season. In arid and semi-arid climates where rainfall is absent during this period, it is important to calculate the amount of water required to leach the accumulated salts from the rootzone.

Assuming that the amount of evaporation can be ignored during the leaching process and that the soil moisture has been restored to field capacity so that the leaching efficiency coefficient f_i equals f_r, the depth of leaching water can be calculated if the leaching coefficients are known. Under the assumption that there is no rainfall, the following equation can be used to calculate the depth of leaching water required:

$$\text{fqt} = W_{fc} \ \text{In} \frac{2EC_{eo} - EC_I}{2EC_{et} - EC_I} \tag{14}$$

where:

f = leaching efficiency coefficient (-);

q = vertical flow rate (mm/d);

t = time required to leach the accumulated salts (d);

Ec_{eo} = salinity in the rootzone at the start of the calculation period (dS/m); and

Ec_{et} = the desired salinity in the rootzone at the end of the period (dS/m).

Impact of Shallow Water Table Management

In the drainage pilot study area, the water table depth is normally high in the winter, crop evapotranspiration is low and irrigation water supplies are limited during the winter season. It is ideal to practise shallow groundwater table management in the winter when wheat is the main crop. Total crop evapotranspiration for wheat is 385 mm. The total seepage inflow during the wheat growing season is 0.5 mm/d * 150 d = 75 mm.

The salinity of the seepage inflow is 5 dS/m and the salinity of the irrigation water is 1.0 dS/m. The water table depth (*WT*) at the start of the season is 0.7 m. The readily available soil moisture for the silty soil as a volume percentage (q) is 0.191. The drainable pore space (m) 0.15. The soil moisture content in the rootzone at field capacity (q_{fc}) is 0.36. It is assumed that all farmers in the area will practice shallow groundwater table management and no additional seepage into the area is induced. Under these assumptions the following steps need to be followed to calculate the change in groundwater table depth and the contribution of the capillary rise ($G = qt$) towards crop evapotranspiration:

— *Step 1.* Calculate z, which is WT - half of the depth of the rootzone (D_{rz}).

— *Step 2.* Estimate the initial equilibrium q for the calculated z. If the change in soil moisture content ((DW) = 0 then q at the start of the next calculation period is equivalent to the equilibrium q for the calculated z. At equilibrium conditions h = z. Otherwise, q = equilibrium q - ((DW/D_{rz}). If q drops below the readily available soil moisture irrigation is required.

— *Step* 3. Estimate the maximum *q*.

— *Step* 4. Calculate maximum G which is maximum *qt*.

— *Step* 5. Check if the maximum G can cover the water uptake by crops. If maximum G minus ET_{crop} is positive then the actual G is equivalent to ET_{crop}. Otherwise, the actual G is equivalent to the maximum G.

— *Step* 6. Calculate the change in soil moisture content (*DW*) = (ET_{crop} - actual G).

— *Step* 7. Calculate the drop in groundwater table (*Dh*) = ((actual G - S_i)/*m*).

— *Step* 8. The *WT* in the succeeding month is the *WT* of the calculated month plus *Dh*.

— *Step* 9. Repeat previous steps for all the months in the calculation period.

The below matrix presents a summary of the calculations of the changes in groundwater table depth and the contribution of the capillary flow towards crop evapotranspiration.

	Dec	*Jan*	*Feb*	*Mar*	*Apr*	*Total*
WT (cm)	70	89	116	173	203	213
D_{rz} (cm)	50	80	100	100	100	
z (cm)	45	49	66	123	153	
q (for *h* = z)	0.42	0.42	0.41	0.36	0.31	
maximum *q* (cm/d)	> 1.0	> 1.0	0.75	0.2	0.1	
maximum *G* (cm)	> 30	> 30	22.5	6	3	> 91.5
S_i (cm)	1.5	1.5	1.5	1.5	1.5	7.5
ET_{crop} (cm)	4.3	5.5	10.1	11.0	7.6	38.5
actual *G* (cm)	4.3	5.5	10.1	6.0	3.0	28.9
DW (cm)	0	0	0	- 5.0	- 4.6	- 9.6

Dh (cm)	19	27	57	30	10	143

The calculation example shows that the water table drops from 70 cm to 213 cm below the soil surface. The soil moisture content decreases towards the end of the growing season but does not get below the readily available soil moisture. Irrigation is therefore not required.

Assuming that the shallow ground-water salinity is equivalent to the salinity of the seepage inflow into the area, the change in salt storage in the rootzone over the winter season is:

$$DS = 289 \text{ mm} * 5 \text{ dS/m} = 1\,445 \text{ ECmm}$$

If the salinity (EC_e) at the start of the season was 2 dS/m, the salinity at the end of the season would be:

$$S_{end} = S_{start} + DS$$

$$S_{end} = 2 * 2 * 360 + 1\,445 = 2\,885 \text{ ECmm}$$

$$EC_{e\ end} = 2\,885/(2 * 360) = 4.0 \text{ dS/m}$$

Equation 19 can be used to calculate the leaching requirements. If it is assumed that the soil salinity should be lowered to 2 dS/m again before the sowing of cotton, the total leaching requirement is:

$$0.9\ qt = 360 \text{In} \frac{2 * 4.0\ 1.0}{2 * 2.0\ 1.0}$$

$$qt = 339 \text{ mm}$$

This leaching might be obtained during normal irrigation during the following season and may not need any special leaching irrigation unless the salts have actually accumulated in the seed zone.

Land Retirement

Land retirement involves taking land out of production because of water shortage and/or soil and water quality